Modern Approaches in Solid Earth Sciences

Volume 18

Background and motivation

Earth Sciences are going through an interesting phase as the traditional disciplinary boundaries are collapsing. Disciplines or sub-disciplines that have been traditionally separated in the past have started interacting more closely, and some new fields have emerged at their interfaces. Disciplinary boundaries between geology, geophysics and geochemistry have become more transparent during the last ten years. Geodesy has developed close interactions with geophysics and geology (tectonics). Specialized research fields, which have been important in development of fundamental expertise, are being interfaced in solving common problems.

In Earth Sciences the term System Earth and, correspondingly, Earth System Science have become overall common denominators. Of this full System Earth, Solid Earth Sciences – predominantly addressing the Inner Earth - constitute a major component, whereas others focus on the Oceans, the Atmosphere, and their interaction. This integrated nature in Solid Earth Sciences can be recognized clearly in the field of Geodynamics. The broad research field of Geodynamics builds on contributions from a wide variety of Earth Science disciplines, encompassing geophysics, geology, geochemistry, and geodesy. Continuing theoretical and numerical advances in seismological methods, new developments in computational science, inverse modelling, and space geodetic methods directed to solid Earth problems, new analytical and experimental methods in geochemistry, geology and materials science have contributed to the investigation of challenging problems in geodynamics. Among these problems are the high-resolution 3D structure and composition of the Earth's interior, the thermal evolution of the Earth on a planetary scale, mantle convection, deformation and dynamics of the lithosphere (including orogeny and basin formation), and landscape evolution through tectonic and surface processes. A characteristic aspect of geodynamic processes is the wide range of spatial and temporal scales involved. An integrated approach to the investigation of geodynamic problems is required to link these scales by incorporating their interactions. Scope and aims of the new series

The book series "Modern Approaches in Solid Earth Sciences" provides an integrated publication outlet for innovative and interdisciplinary approaches to problems and processes in Solid Earth Sciences, including Geodynamics.

It acknowledges the fact that traditionally separate disciplines or sub-disciplines have started interacting more closely, and some new fields have emerged at their interfaces. Disciplinary boundaries between geology, geophysics and geochemistry have become more transparent during the last ten years. Geodesy has developed close interactions with geophysics and geology (tectonics). Specialized research fields (seismic tomography, double difference techniques etc), which have been important in development of fundamental expertise, are being interfaced in solving common problems.

Accepted for inclusion in Scopus.

Prospective authors and/or editors should consult one of the Series Editors or the Springer Contact for more details. Any comments or suggestions for future volumes are welcomed.

More information about this series at http://www.springer.com/series/7377

Eric Lilford • Pietro Guj

Mining Taxation

Reconciling the Interests of Government and Industry

Springer

Eric Lilford
Minerals and Energy Economics
Curtin University
Perth, WA, Australia

Pietro Guj
Centre for Exploration Targeting
University of Western Australia
Crawley, WA, Australia

Responsible Series Editor: F. Pirajno

ISSN 1876-1682 ISSN 1876-1690 (electronic)
Modern Approaches in Solid Earth Sciences
ISBN 978-3-030-49820-7 ISBN 978-3-030-49821-4 (eBook)
https://doi.org/10.1007/978-3-030-49821-4

This Springer imprint is published by the registered company Springer Nature Switzerland AG.
The registered company address is: Gewerbestrasse 11, 6330 Cham, Switzerland

Disclaimer

The opinions and advice offered in this book are general in nature and not specifically directed to the particular situation or needs of any individual country, exploration or mining company, investor or other party. Governments or other entities or persons seeking to improve their taxation policy, legislation and implementation rules and procedures for more effective and efficient collection of mining taxes should consider their particular needs and circumstances and seek further specific advice beyond that offered in this book. Similarly, exploration and mining companies and other entities and individuals interested in improving their taxation outcomes, cost of compliance and rapport with relevant tax authorities should seek further specialised advice specific to their circumstances. The authors cannot accept responsibility for any loss occasioned by any person acting or refraining from action on the basis of material contained in this book.

Contents

About the Authors

Eric Lilford is an Associate Professor of Minerals and Energy Economics in the Western Australian School of Mines: Minerals, Energy and Chemical Engineering faculty of Curtin University. His role includes both research and lecturing at postgraduate level, while holding engineering diplomas and degrees as well as a PhD in mineral economics.

Prior to the commencement of his full-time academic career 3 years ago, Dr. Lilford held positions including Chief Executive Officer and Managing Director of early-stage public mining and exploration companies holding assets and operations across the world, as well as non-executive directorships and chairperson positions of other natural resources, in listed and unlisted public companies. He also has in excess of 15 years' experience in investment banking's corporate finance and specialised finance divisions, 3 years of broking experience associated with the Johannesburg Securities Exchange and over a decade of mining production experience across numerous commodities and in numerous jurisdictions.

Dr. Lilford has consulted to the International Monetary Fund for taxation work revolving around valuations of commodities that are not LME traded, and has consulted to the International Institute for Sustainable Development to provide specialist advice on valuation approaches and methodologies to be used by a specific government for capital gains tax valuation purposes. He also provides minerals and energy economics short courses on behalf of the Department of Foreign Affairs and Trade, Australia, into the African market, and is on the committee for Rare Earths with the Australian Standards Organisation.

Pietro Guj is a Research Professor at the Centre for Exploration Targeting, at the University of Western Australia, and an Adjunct Professor in Mineral Economics at Curtin University's Western Australian School of Mines.

These academic roles were preceded by a distinguished career in the exploration and mining industry, in Asia, Africa and Australia both in industry and government. He held the role of Deputy Director General of the Department of Minerals and

Energy and Executive Director of the Geological Survey of Western Australia after a few years as a finance executive for the Water Authority of WA.

Dr. Guj's main interests are in project evaluation, risk and decision analysis as applied to the minerals industry and in the formulation and administration of internationally competitive resources' regulatory and fiscal regimes, fields in which he has lectured, published and consulted widely internationally.

In recent years, Dr. Guj has been retained by the World Bank to direct research and capacity-building programs designed to improve mining taxation policy and administration frameworks and to address the fiscal challenge of transfer pricing in the context of mineral-rich developing countries, with particular emphasis on Africa. He has also been contracted by the Australian Department of Foreign Affairs and Trade to direct training in the field of mineral economics and governance in cooperation with the University of Witwatersrand in South Africa.

Abbreviations

Au	Gold
APA	Advanced Pricing Agreement
B-BBEE	Broad-Based Black Economic Empowerment
BEE	Black Economic Empowerment
BEPS	Base Erosion and Profit Shifting
CBA	Cost-Benefit Analysis
CbC	Country by country
CGT	Capital gains tax
CFC	Controlled Foreign Company
CFR	Cost and freight
CIF	Cost, insurance and freight
CIT	Corporate income tax
CoG	Cut-off grade
CoW	Contract of work
DBSA	Development Bank of Southern Africa
DCF	Discounted cash flow
DSO	Direct shipment ore
DTA	Double Taxation Agreement
EDI	Extractives Dependence Index
EITI	Extractive Industry Transparency Initiative
FC	Free carry
FDI	Foreign direct investment
FOB	Free on board
FTS	Flow-through share
GDP	Gross domestic product
GST	Goods and services tax
IP	Intellectual property
IDC	Industrial Development Corporation of South Africa
IRR	Internal rate of return
LME	London Metals Exchange

LoM	Life of Mine
LTBR	Long-term bond rate
MIS	Multilateral instrument
MNE	Multinational enterprise
MRRT	Mineral resource rent tax
NGO	Non-governmental organisation
NPI	Non-participative interest
NRS	Net smelter return
NSV	Net smelter value
OECD	Organisation for Economic Co-operation and Development
OTC	Over the counter
PE	Permanent establishment
PIC	Public Investment Commission
PRRT	Petroleum resource rent tax
PSC	Production sharing contract
PV	Present value
R&D	Research and development
ROI	Return on investment
ROM	Run of Mine
RR	Royalty rate
RRT	Resource rent tax
RSPT	Resource super profits tax
SLP	Social and labour plan
SME	Small- to medium-size enterprise
SPE	Special-purpose enterprise
SPV	Special-purpose vehicle
STC	Secondary tax on companies
TIEA	Taxation Information Exchange Agreement
VAT	Value-added tax
WHT	Withholding tax

List of Figures

List of Tables

Chapter 1
Introduction: Context, Objective and Outline of this Book

Abstract This chapter discusses the main issues that arise in terms of the interplay between government and the mining industry in the context of sharing the benefits generated by mining projects under the prevailing mining taxation regimes and summarises the content of the book.

Keywords Mining · Minerals · Government · Industry · Taxation

1.1 Context: Mineral Policy and Governance in the Context of Mining Fiscal Regimes

The exploration, development and exploitation of natural resources opportunities will continue indefinitely. As one ore body or mineral field is depleted, a new one will have to replace it and, in fact, an increase in production will be needed to satisfy growing demand arising from an ever-increasing population and per capita use. Mineral resources and reserves are a dynamic concept because increasing demand will in the long term affect commodity prices, which as a consequence will progressively turn lower grade resources and recycling into commercially feasible sources of supply or encourage substitution.

In most world jurisdictions, mineral resources belong to the state which must manage them to the benefit of its citizens. Past experience indicates that most governments lack the capacity to be effective and, above all, efficient in exploring for, developing and exploiting their mineral resources and equally so in satisfying the related development risk capital needs. As a consequence and despite notable sources of rhetoric to the contrary, nationalisation of the mining industry is at an all-time low even in essentially centrally controlled economies. Governments, particularly in developing countries, must therefore rely on private enterprise to meet the technical and financial challenges and bear the risks inherent in mining.

Governments have a political obligation to be responsive to societal expectations as to how mining developments should be conducted in an environmentally and socially acceptable manner and to the extent to which the related benefits and costs

E. Lilford, P. Guj, *Mining Taxation*, Modern Approaches in Solid Earth Sciences 18,
https://doi.org/10.1007/978-3-030-49821-4_1

should be shared fairly between government and industry. This obligation is reflected in the country's mineral policy with the industry being governed through often complex legislative and regulatory, including fiscal, regimes addressing all facets of the mining cycles.

However, in legislating how the industry should be regulated and, above all taxed, government should keep in mind the need to attract and retain foreign direct investment (FDI) and be cognisant that the mining industry operates in a global context and will tend to direct its capital to those jurisdictions that it perceives would provide the best return on a risk-adjusted basis. In reality there is no simple formula to determine what a 'fair' share of benefits and costs should be and governments will have to position their regulatory regimes to compete internationally with other mineral rich jurisdictions.

Literature on mineral policy and governance generally recognises that mining development generates both monetary and non-monetary benefits and costs. Although the industry tries hard to emphasise the vast range of positive externalities generated by mining projects, the mining taxation dialogue is invariably cast just in monetary terms. The relationship is seen strictly as a zero-sum game balancing government revenues against mining companies' profits, where a gain to one is seen as a loss to the other. Although a symbiotic relationship should exist between government and industry in the formulation of effective but mutually acceptable regulatory regimes, rarely synergistic, win-win arrangements, based on cooperative communication, consultation and awareness of each other's needs and expectations are negotiated. Not surprising, for these reasons industry and government invariably view each other with a degree of suspicion and the relationship is by and large far from cooperative and transparent and in many cases it is downright adversarial.

It is in the spirit of improving this state of affairs that the OECD (2019) has solicited and facilitated the participation of governments, industry and civil society in a Policy Dialogue on Natural Resource-based Development to reach agreement as to the Guiding Principles (Appendix A) on which durable extractive exploration and production contracts should be based and designed to withstand the test of time.

Government is also under continuous pressure from the communities affected by mining projects, which focus primarily on their negative externalities in terms of environmental and social impacts, which they claim are not adequately compensated for by redirecting an adequate proportion of taxation revenues to their region, particularly during periods of mineral commodities boom.

Most governments would argue that industry does not pay nearly enough tax and, in some instances justifiably so. Their argument often states that given any form of fiscal legislation/regulation, industry companies will find a way to minimise or even avoid their tax liability by exploiting loopholes or by other contrived means. While some of these strategies may be, strictly speaking, legal, they are perceived as morally derelict. Governments also allege that industry often engages in tax avoidance practices that may prove to be illegal. Developing countries in particular contend that there is a negotiating power imbalance with industry, as it makes use of highly specialised financial and legal advisors in their negotiations with their departments, which generally lack in internal administrative skill and capacity and can ill afford external assistance. Deliberate misrepresentation by industry of potential benefits to government is also often alleged. In some cases this may prove to be true while in others it may have arisen out of a lack of full understanding on the side of the government negotiators and sometimes of industry of the potential long-term consequence of the terms and conditions agreed to against the background of a very uncertain industry. This is exacerbated in many instances by the inability of government to do anything about it because the relevant conditions may have been locked into 'stability' agreements preventing and/or making any future change extremely difficult.

Requirements for government to be given a free-carried equity interest in project developments and for a mandated level of local domestic equity can also be poorly structured, understood and fraught with legal ambiguity and administrative complexity. Compounding this is a firm but often erroneous belief that through government's holding of a free-carried interest government will over time realise value either as dividend income or by means of capital growth. There is also an often frustrated expectation that, as an equity holder at project level, government and/or domestic equity holders should have the right to appoint a representative to either the Board of the company or at least to the local management committee of the project. Furthermore when such an appointment takes place government views it as a vehicle to gain inside information to further its aims rather than those of the company shareholders at large which does arise conflict of interest issues.

By contrast, industry tends to claim, in some cases with good reason, that government does not fully appreciates how the mining industry really works, the conditions necessary for it to succeed, the risks it takes and in general its capacity to pay. It also claims that mineral royalties and corporate income tax (CIT) are only part of complex taxation packages that includes many other onerous conditions, and that there is little or no consideration of a large number of legislated or voluntary commitments to the establishment of necessary infrastructure and other direct and indirect contributions to the socio-economic wellbeing of the communities and regions in which they operate. Of particular concern is the fact that mining profits vary widely in response to fluctuating commodity prices and that, in good years, this generates a government and popular perception that industry is not paying its 'fair' share, leading to pressure being applied to amend the existing fiscal regime as a matter of urgency. This latter reaction may lead to unilateral, politically motivated knee-jerk reactions and the ultimate enforcement of quite-often counter-productive reforms.

In essence, governments never believe that their level of taxation is excessive, companies rarely agree that they are not paying enough in taxes and the community generally does not consider that the benefits from royalty and tax payments redirected to them are what they should be. Taken from any of the parties' point of view, each of these perceptions can be supported or countered.

That said, and since every mining operation is finite and hence will ultimately be economically depleted, in addition to which a continually growing world's population is noted, it is necessary to consider our "shrinking planet" in terms of minerals and mining. The shrinking planet concept is a metaphor referring to the fact that the world's population continues to grow unabated (Fig. 1.1 below) while occupying a finite global surface/land area, increasing pressures on land and water resources. The utilisation of land and water resources competes with alternative uses and potential benefits beyond mere habitation and survival, such as economic resources extraction. That is, an increasing global population has two significant land impacts:

- it demands larger tracts of land for residential purposes and, more importantly, agricultural purposes in terms of areas to grow crops and keep livestock; and
- as the population increases, the demand for minerals and associated mineral products also increases.

This may result in potential conflict over land use between habitation/agriculture and minerals extraction, processing and fabrication.

Economic theory suggests that the absolute market potential for a commodity, being the highest potential demand for the commodity per person in any individual country, may trend towards the consumptive *per capita* rates observed in first world economies. Specifically, absolute market potential corresponds to the total sales level, by volume or by value that would be realised if every potential user consumed that product, at the optimum frequency rate and at the full rate per use occasion. This theory suggests that many emerging economies, including the notable markets of India and China, have the propensity to increase demand for commodities and metals on a *per capita* basis, and therefore on an absolute basis, as they further develop their economies and their respective populations grow. From Fig. 1.1, China's near-1.5bn population approximates slightly over 19% of the current global population with India's near-1.4bn population marginally below 18% of the global population.

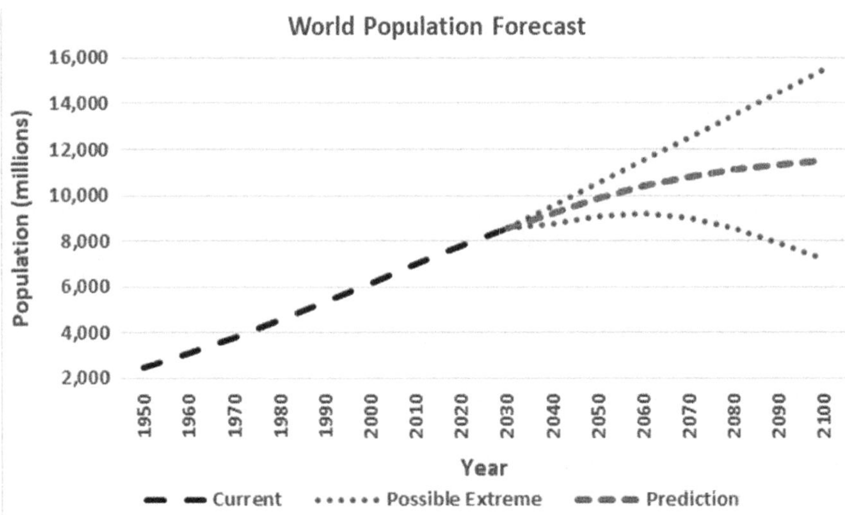

Fig. 1.1 Increasing world population, with forecast (Modified after: https://population.un.org/wpp/Graphs/Probabilistic/POP/TOT/900)

Combining the two countries, their populations are around 37% of the world's current population.

1.2 Objective: Informing and Improving the Government-Industry Tax Dialogue

Many of the world's policy-makers and industry practitioners do not fully grasp the implications of certain actions and impositions on individual orebody developments and, by reflection, on the wider economy. This book aims to clarify some of this uncertainty and dispel some misunderstanding so that ideally both policy-makers and industry can work from a common knowledge base towards greater mutual understanding and hopefully achieve beneficial results from a more symbiotic relationship. It is unlikely that a truly win-win situation will ever arise in the resources industry, but a first step in the right direction would be to avoid lose-lose outcomes, which would be a better result for both sides.

While in the end the outcome of a negotiation will depend largely on the relative bargaining powers of the individual parties, it is paramount that they should fully understand the potential consequence of the terms and conditions of a deal before deciding to accept them. As already discussed, this may not always be the case where it comes to the fiscal regime to apply to a mining project.

One of the reasons for this is because mining projects may have long lives and their performance is fraught with uncertainty, particularly as it concerns commodity prices to which their profitability is highly sensitive. The decision to develop a project is generally highly influenced by its forecast commercial performance based on a 'base case' financial model, and on specific sensitivity and scenario analyses, contained in its feasibility study. In general, both industry's and government's expectations tend to be dominated by the thinking prevalent at the time of the feasibility study, even though they recognise the difficulty of forecasting into the future in realistic terms. Needless to say that as time passes, a project's performance tends to be at variance with plans and, unless the fiscal regime is truly economically efficient or its terms flexible, the proportion of project rent accruing to government and industry respectively may change over time in unpredictable ways. In reality, as discussed in detail later in this book, most mining fiscal regimes are far from economically efficient and the proportion of rent attributable to the various parties tends to change with changes in the economic circumstances of a project, particularly as it regards royalties. As a result, the higher the royalty level the greater the proportion of the geological resources that are 'sterilised' by the need for industry to increase the cut-off grades to maintain their profit margins. Sterilisation of resources reduces the mineable tonnages and consequently the size and/or life of a mining project and generates a corollary of negative economic consequences.

Furthermore, due to an element of distrust, the tendency has been for industry to insist on the fiscal regime not to be subject to future changes by means of 'stability' agreements, irrespective of the nature of unforeseen evolving circumstances affecting either party's interests.

In addition, while the intent of the law may have been clear in the minds of the legislators, its actual wording may have been inadequate to reflect it, creating scope for later ambiguity, misunderstanding and litigation.

The objective of this book is to better inform and improve the government-industry dialogue on tax and other pecuniary issues by systematically reviewing the main components of a typical mining taxation package and other associated imposts and obligations, including provisions for government and community equity participation in projects. It goes on to highlight possible areas of ambiguity and provides quantitative assessments of their impacts on revenue under changing sets of circumstances. The intention is to support opportunities for more open, balanced and cooperative consultation in setting the conditions for future mineral project developments. Sensitivity analyses will be used to test the robustness of projects under changing conditions and to alert decision makers as to the need to retain some elements of flexibility in their arrangements to be able to adapt the applicable fiscal regimes to evolving scenarios and to identify opportunities for win-win arrangements.

The book will also consider how better knowledge may help in streamlining the administration of fiscal collection processes, reducing complaints, objections and litigation and resulting in potential savings in compliance costs for both government and industry.

1.3 An Outline of the Book Content

A close review of mineral taxation principles and prevailing mining taxation packages will be provided. This will include a review of mineral royalties, industry-specific corporate income tax (CIT) provisions and fiscal incentives, other mining industry imposts and obligations including the requirement for domestic and government equity participation in projects. The nature of 'stabilisation' agreements and other key administration issues will also be considered.

Including the Introduction provided in this chapter the book comprises ten chapters.

Chapter 2 deals with the key mining taxation principles and discusses the government challenge of reconciling potentially incompatible economic objectives, such as maximising and stabilising revenue without excessively impacting on the attractiveness of the country as a destination for foreign investment. It deals with the need to establish fiscal regimes that are economically efficient, i.e. where theoretically a change in tax rate does not distort investment decisions, while still keeping in mind the need for revenue stability and administrative transparency and simplicity.

Chapter 3 describes the components of typical mining taxation regime packages, which can combine mineral royalties, corporate income tax, capital gain tax, withholding tax, value-added tax (VAT or GST), import-export duties, excises and a plethora of other minor taxes and imposts.

Chapter 4 deals specifically with the different types of mineral royalties ranging from specific (weight-based), to value- and profit- or rent-based royalties and considers their pros and cons from the points of view of both government and industry. It also consider differences between the royalties regimes generally applied to minerals as contrasted to petroleum production.

Chapter 5 examines corporate income tax provisions and fiscal incentives specifically targeted to the mining industry as a strategy to attract foreign direct investment (FDI). These include tax holidays, immediate write-off or accelerated depreciation of capital expenditure, capital and depletion allowances, and exemptions from VAT, excises, capital gains, import-export and withholding taxes. As further discussed in Chap. 5, while fiscal incentives may be justified in the context of attracting investment, some, if poorly designed and understood, may have deleterious consequences in the longer term in terms of government revenue.

Chapter 6 makes use of the discounted cash flow (DCF) financial model of a realistic gold mine to analyse and compare the distribution of benefits and costs between Government and the mine operator for various taxing instruments including different types of mining royalties, Government's free-carried or participative equity, tax holidays and other fiscal incentives. The model will help identify which trade-offs are more desirable and/or deleterious from the point of view of either party and help achieve feasible compromises while structuring the fiscal regime which would, on balance, be most appropriate for a specific project development.

Chapter 7 deals with Government's requirements for Government's equity participation and for specific levels of domestic shareholding in mining projects, which are typically found in the fiscal packages of developing countries, particularly in Africa. It discusses the difference between participative and free-carried forms of Government's equity and describes how Government legislates to enforce them as well as the requirements for domestic companies' and/or local communities' participation.

Chapter 8 covers the nature and use of 'stabilisation' agreements which are invariably required by the mining industry before they embark into a significant investment in mine development. These agreements are designed to ensure security of tenure and that there will be no changes in the fiscal regime and other mining conditions typically over the life of the mining operations. The chapter discusses how the intent of stability agreements may have generally been clear and transparent at the time of their signature, but how their generally poor legal structures and complex administrative processes may create significant challenges in the long run when the parties are confronted with significant and often totally unforeseen changes in circumstances.

Chapter 9 deals with the challenges of administering a mining fiscal regime in the context of an increasingly globalised mining industry. It contrasts the continuous endeavour of multinational mining companies to adjust their corporate structures in a manner that minimises their taxes at the consolidated corporate level, through cross-border transactions between related parties involving transfer pricing, tax rate arbitrage, treaty shopping etc., with Governments' continuous efforts to protect their tax bases from erosion and profit shifting and to find workable solutions to improve their mining taxation collection and administration frameworks.

Chapter 10 provides general conclusive discussion about the status quo, current trends and desirable potential reform of current mining fiscal regimes. To the extent that these are capable of being accommodated, they would ensure renewed foreign direct investment in mining in developing countries based on Government and industry sharing benefits and costs on the basis of reaching in a less adversarial atmosphere an agreement on the conditions under which a proposed mine development should take place based on a clearer and common understanding of all the related fiscal and administrative issues involved.

Chapter 11 will finally draw some main conclusions and provide general advice as to how a country's fiscal regime and the interface between its tax authorities and the mining industry may be improved towards achieving a more co-operative relationship and easier administration, resulting in a lower level of disputes and compliance cost. The conclusions and ideas are framed separately under the headings of government and of industry.

Reference

OECD (2019) Guiding Principles for Durable Extractive Contracts. http://www.oecd.org/dev/natural-resources.htm

Chapter 2
Mining Taxation Principles and Objectives

Abstract This chapter compares and contrasts the economic objectives and expectation of government and industry and considers to what a degree they can be reconciled and how government's objectives of economic efficiency and revenue maximisation are tempered by the need to compete with other mining-rich jurisdictions to attract internationally mobile foreign direct investment (FDI).

Keywords Economic objectives · Government · Industry · Reconciling differences

2.1 The Role and Contribution of Mining to the Economy, Mineral Policy and Governance

In most countries, natural resources belong to the State which must manage them to the benefit of its citizens. Only seldom does one find that mineral resources in the ground belong to the landowner, an individual, or a private or public non-governmental company.

It is generally accepted that State ownership of natural resources means that all of a country's inhabitants should benefit in some way from their exploitation, although there is growing pressure from the communities directly affected by mining to claim for benefits in excess of adequate compensation for the disturbances arising from mining activities in their territory, in effect for some form of local royalty.

Government and its relevant departments should aim to provide the framework within which resources exploration, development and operations can be conducted in a productive, safe and environmentally acceptable manner, including final land rehabilitation. Many of these frameworks are transcribed into laws, regulations or procedures that dictate the conditions under which a resources development can take place.

With that outline, the primary purposes of minerals policies are to:

- protect and enhance the national interests of the general population by regulating national, natural assets including mineral occurrences;

E. Lilford, P. Guj, *Mining Taxation*, Modern Approaches in Solid Earth Sciences 18,
https://doi.org/10.1007/978-3-030-49821-4_2

- protect mining employees and contractors through labour, including protection of local jobs, health and safety, and other laws;
- ensure government receives its due rents and taxes from the exploitation of natural resources;
- protect the environment, including flora and fauna, from any activity that is recognised as potentially damaging it if left unregulated through protection and rehabilitation laws;
- monitor the activities of exploration and production companies to ensure that they are operating according to the requisite laws and regulations;
- encourage sustainable business practices and operations development;
- prevent illegal activities including those of artisanal miners, unless specifically allowed under prevailing mining laws;
- protect strategic minerals; and
- hold mining executives accountable for their actions.

Some nations, particularly but not exclusively developing ones, may not have adequate domestic technical capacity to successfully and economically exploit mineral deposits to the optimal benefit of their citizens, let alone have access to the necessary capital funds to develop them. Although over the years some countries have opted to nationalise and exploit mineral resources through state-owned enterprises, these attempts have, by and large, proven to be less than successful. Consequently nowadays it is the almost exclusive role of private enterprise to invest in, operate and bear the risk of mineral exploration and mining, while government-owned and operated mining projects are now a rarity. The picture is, however, somewhat different in the case of petroleum where 15 out of 20 largest oil companies are state owned (Cotterelli 2012). National oil companies (or NOCs), however, as discussed in more detail later, invariably secure the services of private enterprise through various contract forms to manage and extract their petroleum resources.

As a consequence, in the case of mining, the role of government has become, in the first instance, to attract the necessary foreign direct investment (FDI) into mineral exploration by creating a perception of their country being highly prospective through the provision of ideally freely available, pre-competitive, geoscientific information.

Although geological prospectivity is the main magnet to investment, government must also create a reputation for low country and sovereign risk. Interestingly many foreign investors will initially overlook these risks and high level of taxation if a country is perceived as highly prospective particularly in the context of securing access to strategic minerals that their own country does not have and desperately needs. As an example, China has a significant strategic exposure to Democratic Republic of Congo (DRC) because it hosts and produces a significant proportion of the world's cobalt, a metal used largely in the aerospace and energy storage industries.

To be an attractive investment destination a country must be seen as endeavouring to facilitate and regulate exploration, development and exploitation of any economically feasible mineral discovery in a safe, socially and environmentally acceptable manner. This will rely on the way it carries out a range of regulatory functions such as maintaining an accurate cadastral system of exploration and mining tenements, allocated in a transparent and equitable manner, managing effective and efficient mining approval processes, and performing the necessary controls to ensure that miners strictly abide to the conditions of their titles, including complying with their safety, environmental, technical and fiscal requirements in a fair and even-handed manner.

The degree to which various mineral-rich countries succeed in creating an attractive perception in terms of mining investment is measured by the annual Frazer Institute Survey using a number of criteria. In their 2018 survey of 83 jurisdictions around the world they rated the state of Nevada as the most attractive jurisdiction in the world for mining investment, followed by Western Australia, Saskatchewan and Quebec (Frazer Institute 2018). Their overall Investment Attractiveness Index combines an assessment of the country's geological prospectivity with its Policy Perception Index. The latter measures the extent to which government policies encouraged or deterred exploration and investment during the year and is subordinate to the country's prospectivity accounting for approximately 40% of a country's investment attractiveness. Policy factors quoted by the Frazer Institute include: 'uncertainty concerning the administration of current regulations, environmental regulations, regulatory duplication, the legal system and taxation regime, uncertainty concerning protected areas and disputed land claims, infrastructure, socioeconomic and community development conditions, trade barriers, political stability, labour regulations, quality of the geological database, security, and labour and skills availability.'

Before becoming a productive mine, however, a project must undergo many years of exploration incorporating sampling, trenching, drilling, assaying and evaluation, requiring significant effort, perseverance and, above all capital. This process, culminating in the completion of a bankable or definitive feasibility study, may typically take between five and ten years with the related expenditure often running into tens of millions of dollars.

Unless the exploration project is owned by a cash flow positive mining company, debt will not be available and the exploration and evaluation phase will generally be funded exclusively through expensive and dilutive equity. Equity is raised on the market, through farm-out and joint ventures or through some form of hybrid debt instrument ultimately convertible into equity.

Companies will update their shareholders and the market after each phase of exploration and evaluation in the hope of satisfying their expectations and of sustaining, ideally increasing, their share price, enabling them to raise additional equity capital as needed in the future on a less-dilutive basis than previously achieved.

The exploration and evaluation phase is also an important period in terms of starting discussions with the government on how the company is to conduct its future mining business in the country and negotiations on any development condition not irrevocably fixed in current laws and regulations. Familiarity with the regulatory regime of a country on the part of company's representatives is essential to identify areas where concessions may be made or taken, or at least discussed. It is equally important for the government officials on the other side of the table to recognise that foreign investment will only be forthcoming if it is perceived as economically worthwhile from the company's point of view to invest there. This involves the negotiating parties being well-versed, besides with local knowledge, with matters of mineral economics, understanding the fundamentals behind:

- commodity markets and expected supply and demands forecasts;
- sources and costs of equity and debt funding and related interest rates (Lilford et al. 2018);
- real and perceived sovereign risk;
- royalty regime and potential impact on operations;
- various additional taxes (import and export duties, withholding taxes, etc.);
- carried or free-carried interests and implications; and
- social and environmental attributes, including local communities, expatriate employees and skills transfer.

In many countries and more noticeably in poorly diversified developing and emerging market economies, the extractive industries are a significant contributor to government revenues (see Fig. 2.1) and as a consequence their economic development is highly dependent on it. Their degree of dependency is measured by the Extractives Dependence Index (EDI) developed by the UNDP (Hailu and Kipgen 2015) which is based on three indicators:

- share of total export earnings from extractive activities;
- revenue from extractive activities as a share of total fiscal revenue which, as shown in Fig. 2.1, can be very high (e.g. 33–88%); and
- value added by extractive activities as a percentage of the total.

According to Davis (1995) countries where mining accounts for more than 8% of GDP are classified as **mineral economies,** while the McKinsey Global Institute puts the threshold at greater than 10% of GDP. Table 2.1, which utilises data sourced from the World Bank Group, lists all the mineral economies the development of which was in 2016 critically dependent on various sources of mining-related benefits.

As summarised in Table 2.2 below, mining benefits can be either monetary/financial, or non-monetary/economic.

The first group includes primarily mineral royalties and corporate income tax (CIT), and a variety of other, individually minor but collectively significant, government taxes and imposts directly levied on mining projects. It may also include dividends from equity in mining projects, that many governments in developing countries retain as either a non-participative (i.e. free-carried) interest, or alternatively as a participative one.

In addition to being a direct employer, the resources industry is also a developer of communities through education and training to satisfy its necessary skills, both specific to the mining industry, but also deployable in other sectors of the economy. As a consequence, the mining industry generates significant indirect monetary/

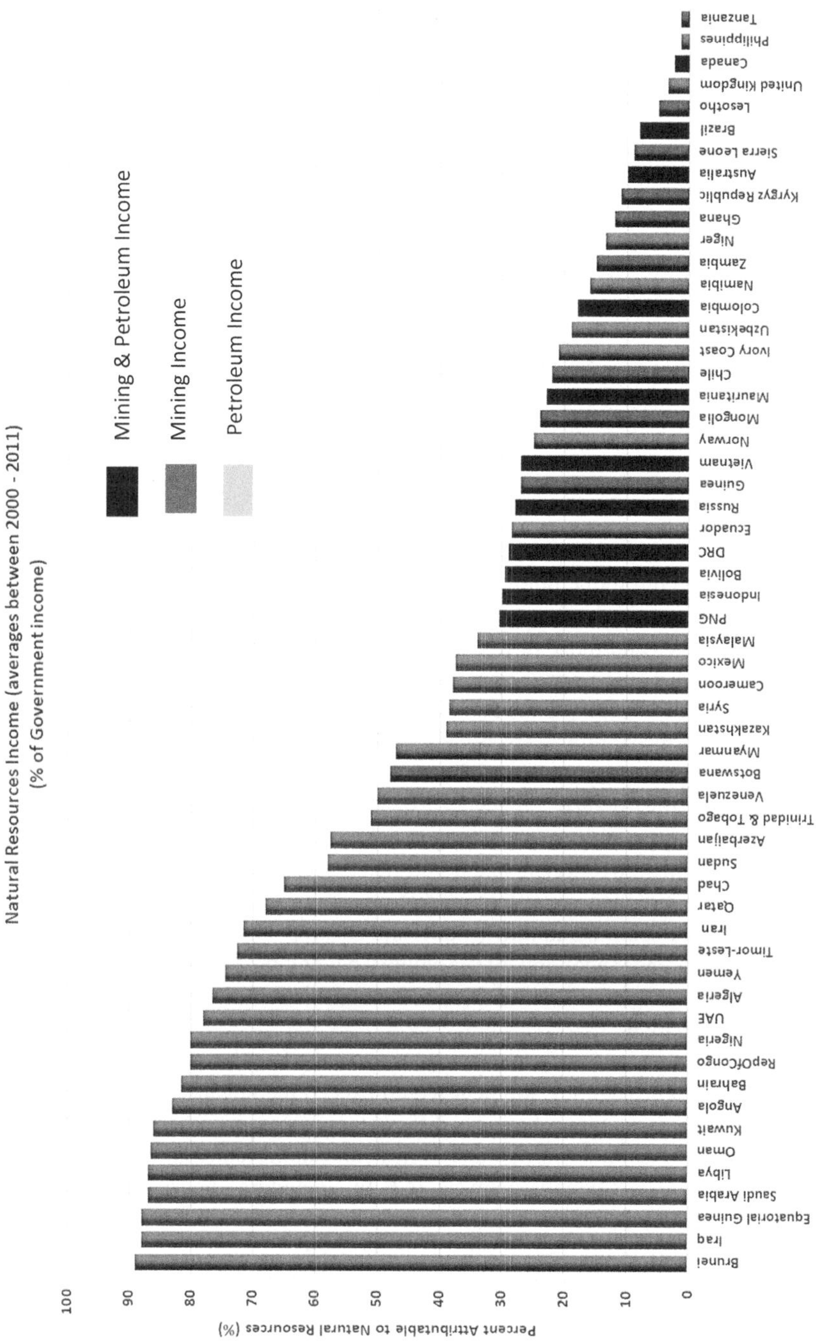

Fig. 2.1 Receipt from natural resources as a percentage of government income (Compiled from IMF Fiscal Affairs Department data)

Table 2.1 List of mineral economies as of 2016 (Data source World Bank Group)

Mining as a percentage of GDP	
Country name	%
Burkina Faso	10.25
Chile	9.82
Dem. Rep. Congo	13.18
Guinea	9.76
Guyana	11.78
Kyrgyz Republic	8.10
Liberia	17.58
Mali	8.97
Mongolia	17.17
Mauritania	22.28
Papua New Guinea	11.33
Sierra Leone	9.89
Suriname	24.00
Togo	11.93
Zambia	10.88

financial benefits through employment and other multipliers contributing additional income to the fiscus in the form of the personal and corporate income taxes paid by both its employees and those of its service providers and suppliers.

The second, non-monetary/economic group includes a potentially vast range of obligations directly imposed on the mining company as conditions of its mining permit. Some of these, referred to by some authors as 'quasi royalties' and listed in Table 2.2, may be very onerous. This category may include establishment of common use infrastructure, payments to local communities in excess of actual disturbance, preferential local procurement at non-market competitive rates, satisfying and maintaining government and domestic non-government equity level requirements free or at less than prevailing market prices etc.

Non-monetary socio-economic benefits and costs are also likely to arise indirectly throughout the broader economy of the country beyond the immediate area impacted by the project. Economists refer to these as economic externalities which can be both positive and negative.

Community opinions may be divided as to whether the positive externalities associated with a project development would exceed the negative ones thus justifying approval. This happens because different opinions may be based on vastly different value bases and the methodologies for valuing externalities in monetary terms for use in cost-benefit analysis (CBA), including the so-called 'shadow pricing' and 'social discount,' while theoretically rigorous, are far from unambiguous and universally politically acceptable in practice.

It may, for instance, be argued that the development of mineral deposits located in areas where infrastructure is either scant or non-existent facilitates and assists in regional socio-economic development. This is because establishment of critical infrastructure developed for a mine, which typically includes road access, electricity

Table 2.2 Characterization of mining benefits and costs

	Monetary/financial	Non-monetary/economic
Direct	• Mineral royalties • Corporate income tax (CIT) • Capital gain tax (CGT) • Dividends from free-carried government equity • Export-import duties and excises • Tenement fees and rentals • Petroleum signature/production bonus • Petroleum resource rent tax (PRRT) • Petroleum production sharing • Withholding tax on cross-border dividends, interest and other remittances. • Various other taxes and imposts, e.g. VAT, stamp duty, payroll tax, and fringe benefit tax if paid on behalf of employees	• Contribution to the cost of establishing or upgrading common-use infrastructure • Provision of utility services to near mine communities • Compensation payments to the community in excess of actual disturbance • Obligation to provide local training and employment • Requirements to establish government and non-government domestic equity ownership at favourable terms • Preferential procurement from local suppliers
Indirect	Effect of economy-wide employment and other multipliers: • Personal income tax levied from the employees of: – mining companies and of – their service providers and suppliers • Fringe benefit tax on the value of accommodation, meals etc. provided for free or subsidised by the mining company • Corporate income tax and other taxes and imposts levied from service provider and supplier companies • Taxation of shareholders' dividends	Socio-economic externalities: • Positive: – Regional development of frontier areas – Contribution to regional geoscience knowledge and investment attraction – Contribution to the country's strategic resource self-sufficiency • Negative: – Impact on the natural environment – Impact on indigenous communities – General 'resources curse' effects

(grid power or local power), potable and industrial water, accommodation, medical facilities, potentially a landing strip, rail and other capital assets, will facilitate development of other, potentially beneficial, commercial enterprises in the region.

Others, however, may argue otherwise. For instance, a major concern for the Gabonese government surrounding the exploration and development of their vast potash resources and on-shore oil reserves is the easier use by local groups of the potential new roads to access parks and reserves to extend their poaching reach.

Accepting that mineral occurrences are finite and that the exploitation of a specific orebody is conducted on a depleting basis, the issue also arises as to how to ensure that the wider national socio-economic benefits from mining-related revenues from royalties, corporate income tax and all other forms of duties, tariffs and imposts, are invested in a manner that makes them sustainable for future generations. While the question of how these receipts should be deployed for the benefit of the country and hence the overall economy is beyond the scope of this book, appropriate use of this income should include building new and maintaining existing infrastructure, repaying local and foreign debt, facilitating discovery of new reserves for future exploitation and reinvesting into gathering regional geoscientific information to promote the country's prospectivity and to attract additional investment in the sector.

There is, however, a wide body of opinion that argues that, based on past experience in some mineral-rich developing countries, appropriate use of mining revenues has not occurred and that, far from contributing to social wellbeing, has resulted in the related negative externalities overcoming the positive ones with the community being left worse off. This effect, referred to by a number of authors as the so-called 'resources curse' (Kronenberg 2002). Whereby the country hosting economically attractive endowments of non-renewable minerals may experience stagnant economic growth or possibly even economic contraction, despite their mineral wealth. Asides from squandering of mineral revenues by incompetent and/or corrupt governments, the country may be focussing all or most of its efforts and allocate significant resources to the 'leading' minerals industry at the expense of or neglecting other potentially 'lagging' economic sectors. As a consequence, the country's economic wellbeing becomes largely dependent on commodity prices and associated price volatilities, generating an equally volatile economy. With excessive focus on the minerals industry, the evolvement of the 'Dutch Disease' (Van Wijnbergen 1984) may arise where, due to the attraction of labour to that industry, skilled workers in other sectors or industries relocate to the minerals industry, chasing higher wages and the promise of increased wealth. Their relocation creates a void in the lagging industries from which they came, causing those industries to suffer and potentially collapse. The end result is that a once diversified economy experiences reduced diversification, exposing the entire economy to the vagaries of the minerals markets.

Although the industry tries hard to emphasise the vast range of positive externalities generated by mining projects, the mining taxation dialogue is invariably cast almost entirely in monetary terms. The relationship is seen strictly as a zero-sum game balancing government revenues against mining companies' profits, where a gain to one is a loss to the other. Thus, although a symbiotic relationship should exist between government and industry in the formulation of effective and mutually acceptable regulatory regimes, rarely the possibility of synergistic, win-win arrangements, based on cooperative communication, consultation and awareness of each other's needs and expectations actually occurs. Not surprising, for these reasons industry and government invariably view each other with a degree of suspicion and the relationship in most jurisdictions is by and large far from cooperative and transparent and in many cases it is downright adversarial.

2.2 Government's Fiscal Objectives versus Corporate Commercial Objectives

Tilton and Guzman (2016) state that government's economic policies and objectives should aim to "...maximize society's welfare" by maximising net economic benefit to the benefit of its citizens, compared with the private enterprise's objective of maximizing value to the benefit of its shareholders. This should encompass all monetary and non-monetary, tangible and intangible commercial, socio-cultural, and environmental values less their corresponding costs.

Unfortunately many governments fall short of their mandate to maximize society's welfare, frequently because of the short-term focus of the political process, predominant emphasis on monetary benefits relative to longer-term economic ones, pervasive corruption and cronyism, and, in many cases, because of sheer incompetence. On the business side, failure to maximize profits is more often than not explained by market forces negatively affecting their profitability. This is not to say that private companies are immune from corruption, cronyism and incompetence.

Government's economic policy is implemented through two distinct mechanisms: fiscal policy, which focuses on taxation levels and government spending, and monetary policy, which focuses on the central bank functions of setting money supply and interest rates. The interplay of these with overall government budgets determines whether the economy will perform on the basis of generating an annual surplus or, more often than not a deficit.

As already pointed out, when it comes to setting a taxation regime for mining, debate tends to gravitate primarily on monetary values. Taxes and other imposts payments levied from the mining industry generally accrue to 'unallocated consolidated revenue,' and are then redistributed according to government budgetary priorities reflecting the prevailing political imperatives, national interest and the need to create stimuli for local and foreign investment to secure further economic development and growth. In spite of growing pressures, only rarely are mining revenues hypothecated exclusively for development in the regions immediately affected by the mining operations as, in the absence of horizontal fiscal equalisation policies, this would result in pockets of unequal wealth being created in different parts of a country. In general Government spending tends to focus on:

- social security and welfare (aged, veterans, disability, families with children, sickness and unemployment);
- health (medical services, pharmaceutical, public hospitals, health services, administration);
- education (higher education, vocational, government and private schools, student assistance);
- defence (military and police);
- public services (legislative and executive affairs, financial and fiscal affairs, foreign affairs, research, superannuation (pension));

- public order and safety (courts and legal services);
- amenities (housing, community, recreation);
- fuel and energy; and
- other (agriculture, forestry, fishing, mining, manufacturing, transport and communications).

As minerals extraction is not a sustainable activity, it is imperative that some of the income generated and delivered to the government's treasury sourced from minerals-extraction activities be used wisely to ensure that alternative, sustainable, economic activities are developed for the continuing benefit of future generation. An excellent example of intergenerational equity is the development of the tourist industry in Dubai, United Arab Emirates, from the proceeds of a once-flourishing, income generating oil industry.

Broadly, governments and their administrative bodies typically regulate the labour market, national ownership and legislate other areas of intervention and aspects impacting on national interest.

The value maximisation objectives of private enterprise, by contrast, are unambiguously monetary. In practice, maximising value consists in managing the company's investments and operations so as to achieve long-term profitability, solvency and liquidity, growth through value-adding and sustainability to counteract reserves depletion. In addition to be in a position to raise the necessary development capital, private enterprise must provide investors/shareholders with suitable returns commensurate with their risk exposure.

Corporate objectives will outline how, given a specific regulatory regime, this is to be achieved by managing those components of the business that can be influenced or controlled including optimisation of production rates, targeted grades, operating costs, capital programmes, sales terms, etc. By contrast, commodity prices and exchange rates represent exogenous risks which cannot be controlled by the business, but risk management strategies, including hedging and forward sales, may be adopted to mitigate their potential effect.

Fortunately in many instances the objectives of both government and business overlap and combine hopefully resulting in synergy particularly in the areas of:

- employment—individual and family income, with individual taxes payable to government, contributions to retirement and medical benefit funds;
- skills development—further enhancement of existing skills or upskilling (acquiring new skills) that may be used elsewhere;
- community and regional development;
- social development; and
- infrastructural development.

In the final analysis, when it comes to setting and/or accepting a fiscal regime, both government and corporate enterprise aim to maximise income. The former must be conscious of the need not to harm and/or discourage business investment, that is to say of not to kill the golden goose, while the latter must resign itself to the inevitability of having to pay taxes and potentially of having to make socio-economic contributions at the regional level.

2.2.1 Investment Attraction

An important consideration is the role of a government as it impacts on a country's attractiveness for foreign direct investment (FDI) and its effect on the overall economy. A government's role should, ideally, be neither an idle one nor an extreme interventionist one. In the context of the resource industry, Tilton and Guzman (2016) contend that governments must provide the regulatory framework in which exploration and mining activities may occur, while simultaneously leaving the actual function of mining and beneficiation to the expertise and abilities of private companies. The regulatory framework managed by the governing bodies incorporates the various laws, regulations and rules dictating corporate and civil behavior as well as comprehensive laws and rules around environmental, social and fiscal behavior.

Accepting that FDI in exploration and mine development is globally mobile, governments need to be conscious of having to compete for it internationally with other mineral-rich jurisdictions. To be successful a country needs to develop its mineral policy and regulatory regime with cognisance of the main criteria that make a jurisdiction attractive to FDI (Guj et al. 2013). These include a perception of the country as having:

- high geological prospectivity;
- impartial rule of law based on a transparent legal and fiscal framework that is:
  - simple and unambiguous,
  - fair and equitable,
  - stable and predictable and ensuring security of tenure, and
  - easy and efficient to comply with;
- political stability; and
- attractive logistics, infrastructure and in-country skills availability.

As already discussed in Sect. 2.1, the perception of geological prospectivity, particularly if supported by a history of mineral discovery and successful mine development, is by far the greatest magnet to FDI. Companies may explore in a highly prospective country even if its fiscal regimes is not highly competitive. In general investors will tend to put up with higher levels of taxation if tenure and the fiscal system are perceived as stable or can be stabilised by means of a 'stability' agreement.

Stability agreements are generally insisted upon by industry where a country, which is relatively new to resources development, attempts to attract FDI for new resources projects by providing generous fiscal and other incentives. These incentives are justifiable against the significant benefits that the overall development of a new industry with associated infrastructure and skills development and transfer will provide to the country. However, as discussed in more detail later in this book, the long-term consequence of some of these incentives are often not fully appreciated by government at the time of signature and as the industry matures the initial benefits provided by the early developers tend to be forgotten generating resentment later in the life of projects.

2.2.2 Economic Efficiency and Equity

Accepting that mining depletes an orebody, government must recognise that the mining activity itself is not sustainable at the individual project level. The fact that an orebody can only be exploited once also places the onus on government to ensure that each orebody be optimally depleted. This in turn means that the rules and regulations that it imposes on mining companies should impact as little as possible on its operating decisions particularly as it concerns determining the commercial viability of mineable reserves and grades. That is not to say that governments should not levy royalties, taxes, duties or other income-generating imposts on mining operations, but rather that these imposts should not severely alter the manner in which ore-depletion is to be executed relative to what it would have been in their absence.

A mining tax system where a change in the tax rate does not alter investment and/or operational decisions is said to be economically efficient. Such a system could, in theory, only exist in an ideal world of frictionless markets and perfect information. Under this set of circumstances investors could expect a 'normal' rate of profit adequate to justify commitment of their capital to individual mining projects, while government could appropriate any profit in excess of the normal profit or economic rent (Hartman and Guj 2013), the so-called 'super-profit,' without, in theory, distorting the company's investment decisions. The nature and application of resource rent based taxes will be discussed in more detail in Chap. 4.

At a macroeconomic level economic efficiency can be viewed in terms of its:

- 'Productive Efficiency' when the output of the economy is being produced at the lowest cost;
- 'Allocative Efficiency' when resources are being re-allocated to their most productive use to generate the ever changing mix of goods and services that society requires; and
- 'Distributional Efficiency' when the composition of the output is such that consumers would not wish, given their disposable income and market prices, to spend it in any different way.

In reality, markets and information in the context of mining are far from perfect and most mining tax instruments in current use are to a lesser or larger degree economically inefficient. As a consequence an increase in the tax rate generally causes a mine to react by increasing the cut-off grade, that is to say the lowest grade at which ore is worth mining relative to mining costs, thus increasing the unit value

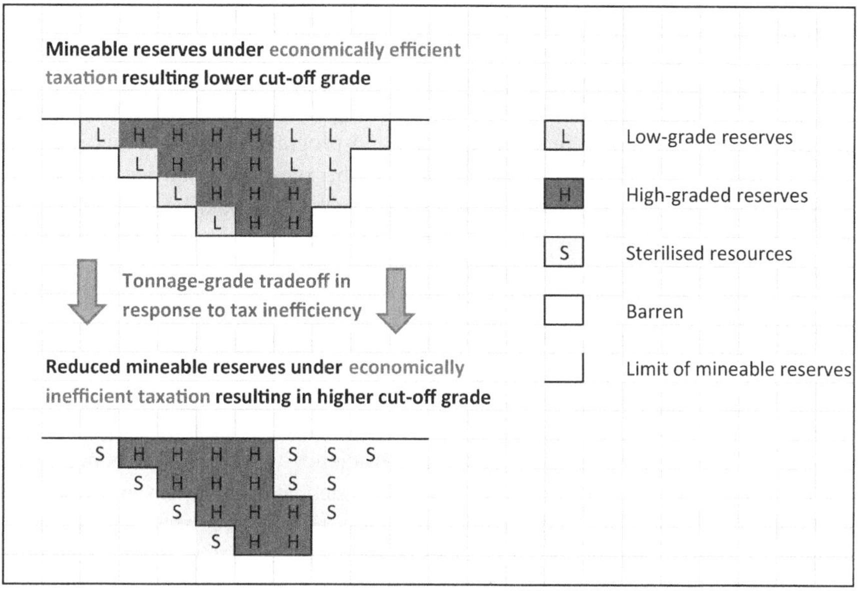

Fig. 2.2 Diagram schematising 'high-grading,' i.e. how mineable reserves decrease from larger tonnages at lower grade to lower tonnages at higher grade as an effect of economically inefficient taxation

of production and maintaining its profit margins, as schematised in Fig. 2.2. This practice of 'high-grading' an orebody, which intuitively sterilises some of the resources and reduces the life of a mining operation, in response to the use of economically inefficient imposts, such as volume or weight-based royalties, has been analysed by Lockner (1965), Steele (1967), Laing (1976) and Gillis et al. (1977). A dissenting view is offered by Burness (1976) who argues that the life of an operating mine remains unaffected by a royalty,

These issues are further elaborated in Chap. 4 which provides a detailed discussion on the various types of royalties and modelled in detail in Chap. 6 which demonstrates and quantifies the impacts that unit-based or value based royalties have on the State and a resources company making use of a Case Study based on an Australian Gold mine.

In practice, however, it is not unusual to schedule production in a manner that deliberately exploits lower grade material in high-commodity-price periods and to mine higher grade material when/if prices drop. While this strategy may maximize the life of the mine, which naturally benefits many for longer, it does not necessarily maximize its profits, value or returns.

It is also not uncommon for miners to allow the extraction and stockpiling of sub-economic material, i.e. with grades lower than the current cut-off grade, particularly if this facilitates production scheduling. Currently uneconomic, lower-grade material may come useful for later blending with anomalously high-grade ore to maintain stable mill feed quality. Stockpiling of lower-grade material may also be justifiable if the price of the commodity is highly volatile and there is an expectation that it may become economic over time if/when a price rise eventuates. Under these circumstances the sub-economic material is said to have 'real option' value.

In a tax context, equity, that is to say ensuring a just and fair treatment for all taxpayers, is often defined in a proportional and progressive context that is to say reflecting different taxpayers' 'ability-to-pay.' In this sense an economically efficient tax system would also be equitable in that the level of tax paid by more profitable producers of higher grade ore of more valuable commodities in greater proximity to markets, would be proportionately higher.

At an individual project level, however, truly economic optimisation is often difficult to achieve in practice because, even though the economic benefits generated by a project may exceed its costs, its development decision may generate losers among some sectors of the community and other stakeholders affected by it and as a consequence political pressure against its development. In this light Kaldor (1939) and Hicks (1939) have redefined 'economic efficiency' in terms of a change being economically efficient if those who gain could adequately compensate those who lose and still be better off. In practice the process of transferring benefits from winners to losers, such that no one would be worse off than they would have been in the absence of the project, may prove to be itself rather fraught with difficulties and inefficient and, at times, impossible. Much of the difficulty arises out of generating comparability between the measures of benefits and costs where it come to valuing some externalities by means of 'shadow pricing' and the irreconcilable differences in values that may be placed on some of them by different parties with firmly held different beliefs, cultural and other value bases.

2.2.3 Revenue Maximisation and Stability and Sharing Benefits and Costs Between Industry and Government

As discussed in Sect. 2.2.1 and notwithstanding theoretical arguments about government being capable of appropriating the lion's share of super-profits with impunity, the globally competitive nature of attracting FDI imposes some limitations as to the proportion of the rent that governments can appropriate before investing in their jurisdictions become unattractive. It is estimated (EY 2019) that a jurisdiction imposing an increase of 10% in its corporate income tax rate will incur a 7.5% decrease in foreign direct investment (FDI) in the country (PwC 2018). Nevertheless, government revenues from mining activities in mineral-rich countries may be very significant. As an example in 2017–18 Australia collected a total of around $ 31 billion in mining taxes including $ 18.6 billion in corporate income tax and $ 12.0 billion in mineral royalties. The policy dilemma becomes how to maximise revenue collection without deterring FDI, which ultimately means collecting what the 'market will bear' rather than what government or the people may consider as a 'fair' level of mining taxation.

It is true that, once discovered, natural resources are captive to a country and that, in theory, as later discussed in Chap. 8, in the absence of a stability agreement, government could increase tax rates and appropriate an increasingly larger proportion of the rent generated by their exploitation. However, it is also true that this would discourage further investment in the country deferring further exploration and the discovery of new orebodies, thus stalling any growth in the industry in their country. In the final analysis jurisdictions need to decide how to manage their tax base, in effect decide whether they wish to have fewer existing mines highly taxed or more mines more lightly taxed but providing avenues for continuing vigorous socio-economic development and growth.

Unfortunately cyclical increases in commodity prices, as discussed and modelled in detail in Chap. 5, tend to leverage company profits more than government revenues. This creates a popular perception that mining does not pay a 'fair' share of taxes. Not surprisingly this generates political pressure for changes in the tax laws to capture more of the economic rent, leading, more often than not, to knee-jerk reactions in terms of poorly conceived and counterproductive fiscal reform.

Unless the economy of a country is highly diversified, another major issue from the government's point of view, particularly in mineral-dependent economies in which forecast mining revenue represents a large component of their budgets, is that revenue is inherently unstable because of the high volatility of mineral commodity prices. While revenue stability should be given priority when designing the mining tax regime in mineral-dependent economies, reality is that taxation instruments that ensure revenue stability tend to be highly economically inefficient and inequitable.

While a large body of literature is dedicated to possible strategies to overcome the revenue instability effect of the boom and bust cycles in the mining industry, primarily through income equalisation schemes such sovereign funds, the degree to which these are being established is relatively low. This is particularly, but not exclusively, the case in poorer countries where the political system tend to use all available funds to satisfy critical immediate and in some case politically inspired short-term needs.

Beyond taxes, royalties and a variety of other imposts, as will be discussed in Chap. 7 in detail, some governments and notably those of emerging market economies, require holding a free-carried interest in local operations. While this may seem to be a guaranteed way in which government could receive income potentially through a dividend on its equity stake, this perception may be misjudged, particularly in the short-term. It is highly likely that, if a company has determined that it is still attractive to invest in a country despite the free-carried imposition it may decide to reinvest its profits in new capital assets whenever they are generated to achieve capital growth rather than paying any dividends at all. To the extent that government is generally precluded from trading or disposing of its shares on the market, more often than not it lends not deriving any monetary benefits from holding a free-carried equity in a mining company either in the form of dividends or capital gain in spite of the financial successes of its operation. The alternative adopted by some government of either requiring payment of dividends irrespective of a company's emerging capital needs or, even worse, as a percentage of annual net cash flows has often the effect of rising the company's cost of capital thus reducing its profitability.

Often the mining company will need to establish specific transport and services infrastructure in order to gain access to remote project areas to develop, construct and carry out mining activities, and to provide an export route for its product. This infrastructure can typically also be used by the local population, especially if it includes roads, rail, water reticulation and power. It is then logical then to think that, once mining has terminated after depletion of the orebody, this infrastructure will continue to generate benefits through its future use by the local population. This adds to the sustainability appeal typically provided by the mining industry.

2.2.4 Transparency, Simplicity and Ease of Administration

Once the mining operation is up and running, government must keep a close watch to ensure that the marketing and sales of the product occur in a manner that prevents mining operations from shifting taxable income offshore prior to adequate taxes being correctly calculated and paid in the country hosting the mine. This type of tax minimisation and avoidance issues involving the transfer price of goods and services in cross-border transactions between related parties will be discussed in greater detail in Chap. 9.

In an endeavour to become internationally competitive the majority of mineral-rich jurisdictions have in recent years reformed or redrafted their mining tax policy, legislation and related administrative regulatory frameworks. Their fiscal reform was inspired by the legislation of countries featuring a well-established and successful mining sector like Australia, Canada and the USA. The objective was to achieve regulatory frameworks that are clear, unambiguous and transparent and therefore easy to explain and justify to both taxpayers and citizens, as well as harder to manipulate thus reducing the likelihood of abuse and taxpayer disputes.

Another objective should be for tax rules to be as simple as possible to facilitate their administration and reduce the cost of compliance for both government and industry. While everybody would agree that simplicity is highly desirable, in practice it is very hard to achieve as the legal system inevitably tends to increase in complexity over time because of continuous amendments to meet emerging circumstances and evolving socio-economic expectations, enhancing FDI attractiveness and continuing to accommodate past irrevocable legacy decisions.

Developments may take place either in compliance with the standard requirements of the prevailing mining legislation or, as it is most frequently the case in developing countries, under the unique terms and conditions of often confidential, special (stability) agreements drafted for individual specific projects. Needless to say that the co-existence of multiple and diverse administrative regimes imposes, as will be discussed in more detail in Chap. 9, an unbearably heavy administrative load on government. The combination of all these factors leads to policy inconsistencies and makes the process increasingly opaque and difficult for civil society, the media, and parliament to monitor mining revenue collection. Unfortunately the legislative amendments and reform that would be needed to simplify fiscal systems and make them more transparent are hard to implement because of the rigidity of some of the mining agreements, as discussed in more detail in Chap. 8, and the resistance and political clout of some of the main mining companies.

The issue of transparency has acquired significant prominence in recent years through initiatives undertaken by the OECD and the Extractives Industry Transparency Initiative (EITI). As stated throughout this book, the natural resources occurring in any country belong to its citizens and therefore benefits accruing from the exploitation of those resources must be shared with the populace as a whole. The EITI established an international set of standards to promote open and accountable management of oil, gas and mineral resources, focussing on the voluntary disclosure of information throughout the extractive industry value chain, from how exploitation rights are applied for and awarded, through to how revenues are charged and received by the government and how this fiscal income benefits the public.

Through providing best practice principles, the EITI's aim is to strengthen public and corporate governance, to promote understanding of natural resource management and to provide the data to inform reforms for greater transparency and accountability in the extractives sector. There are approximately fifty-two countries that have or are in the process of implementing the EITI principles, and in each of them the EITI is supported by a coalition of government, operating companies and civil society at large (https://eiti.org/homepage).

2.3 Reconciling Potentially Conflicting Government Objectives

Some of the government objectives, considered in Sect. 2.2, are mutually incompatible and as a result cannot be optimised simultaneously. As a consequence, most mining taxation regimes are based on compromises that, while not optimal in any single respect, achieve an acceptable blend of the various economic, political and developmental outcomes considered desirable in the context of individual countries.

For instance, maximising revenue would discourage FDI, reduce or defer mineral discoveries and future mine developments affecting industry sustainability and ultimately growth of the tax base. On the other hand, providing excessive and poorly designed fiscal incentives would reduce revenue and make it unpredictable.

Pursuing economic efficiency and equity would lower revenue stability and lead to more complex administration. Furthermore, maintaining regime stability and investor certainty, particularly if involving the use of 'stability' agreements, would result in the coexistence of multiple tax systems, leading to inefficiency, inequity among different projects, lower revenue and administrative complexity.

Take Away Points

Government's point of view

- Consider and, if possible quantify, both the monetary and non-monetary, direct and indirect benefits and costs when assessing the merit of a mining project proposal. Do not focus exclusively on short-term revenue, political and budgetary pressures.
- Adopt a longer-term planning approach by directing some of the income generated by unsustainable minerals-extraction activities to ensure that alternative, sustainable, economic activities are developed for the continuing benefit of future generations.
- To attract FDI create a perception that the country has high geological prospectivity, impartial rule of law, political stability; and attractive logistics and infrastructure.
- Depending on the country's economic dependence on the mining industry attempt to reach a workable compromise between the incompatible objectives of economic efficiency and that of revenue maximisation and stability, while also focusing on equity and ease of administration.

Mining industry's point of view

- Emphasise, document and widely publicise besides the revenue from taxation, the socio-economic benefits and multipliers generated by the project.
- Ensure the support of local communities by aligning their interests with those of the company through local employment and procurement and other tangible and visible community benefits.
- Carefully consider the attractiveness of the geological setting, regulatory and fiscal frameworks, infrastructure and the country's sovereign risk ranking before investing in it. If a decision to invest is reached, adjust the risk-premium in the company's return expectations accordingly and seek, if possible, a stability agreement.

References

Burness HS (1976) On the taxation of nonreplenishable natural resources. J Environ Econ Manag 3 (4):289–311

Cotterelli C (2012) Fiscal regimes for extractive industries: design and implementation. IMF, Fiscal Affair Department

Davis G (1995) Learning to love the Dutch disease: evidence from the mineral economies. World Dev 23(10):1765–1779

EY 2019 (2019) Worldwide corporate tax guide. https://www.ey.com/gl/en/services/tax/worldwide-corporate-tax-guide%2D%2D-country-list

Frazer Institute (2018) Annual Survey of Mining Companies. https://www.fraserinstitute.org/categories/mining

Gillis SM et al (1977) The Indonesian mining sector: tax and related policies. HIID, Cambridge

Guj P, Bocoum B, Limerick J, Meaton M, Maybee B (2013) How to improve mining tax administration and collection frameworks: a source book. World Bank/CET, Washington, DC

Hailu D, Kipgen C (2015) The extractive dependence index (EDI). United Nations Development Program, New York, NY

Hartman F, Guj P (2013) Chapter 12. Mineral taxation and royalties. In: Maxwell P, Guj P (eds) Mineral economics: Australian and global perspectives, 2nd edn. Monograph N. 29 of the Australasian Institute of Mining and Metallurgy - BPA Digital Burwood, Victoria, Australia

Hicks J (1939) The Foundations of Welfare Economics. Econ J 49(196):696–712. https://doi.org/10.2307/2225023

Kaldor N (1939) Welfare propositions in economics and interpersonal comparisons of utility. Econ J 49(195):549–552. https://doi.org/10.2307/2224835

Kronenberg T (2002) The Curse of Natural Resources in the Transition Economies. Working Papers241, Leibniz Institut für Ost- und Südosteuropaforschung (Institute for East and Southeast European Studies).

Laing G (1976) An analysis of the effects of state taxation of the mining industry in the Rocky Mountain states. Masters thesis. Colorado School of Mines

Lilford EV, Maybe B, Packey D (2018) Cost of capital and discount rates in cash flow valuations for resources projects. Res Policy 59:525–531. https://doi.org/10.1016/j.resourpol.2018.09.08

Lockner AO (1965) The economic effect of the severance on the decisions of the mining firm. Nat Resour J 4:468–485

PWC (2018) Battle of the taxes. www.pwc.au/africadesk

Steele H (1967) Natural resource taxation: resource allocation and distribution implications. Gaffney:233–267

Tilton JE, Guzman JI (2016) Mineral economics and policy. RFF Press. Resources for the Future, New York

Van Wijnbergen S (1984) The 'Dutch Disease': a disease after all? Econ J 94(373):41–55. https://doi.org/10.2307/2232214

Chapter 3
Components of a Mining Taxation Package

Abstract This chapter considers the various tax instruments frequently found in mining taxation packages. It describes the main characteristics of different types of mineral royalties and how they are combined with corporate income tax and other mining-specific imposts in a manner ideally designed to secure that an adequate share of benefits from mining activities are enjoyed by the whole country population.

Keywords Tax instruments · Mining taxation · Mineral royalties · Economic impacts

3.1 General Considerations

Asides from a plethora of other imposts, mineral royalties and corporate income tax (CIT) represent the main sources of government revenue in the context of mining. Industry, by contrast, will adhere within the legal constraints, to the popular principle of "paying the least and the latest." The ever increasing financial planning sophistication associated with the resources sector means that taxable entities, whether companies owning and operating individual mining projects or their holding companies, manage to structure and report their taxable incomes in ways that minimize their CIT payments at either the operational project level or at the consolidated group level. While tax minimization is essentially a legal and normal practice, the consequence is generally that mining companies pay less corporate income tax than expected by governments' treasury departments and their taxation authorities.

Optimal tax structuring of any operation is a legitimate managerial priority in creating shareholders' value as long as it is effected within the ambit of prevailing law. Stated differently, where tax authorities have provided the rules around which taxes are calculated, it is incumbent on operators within that tax jurisdiction to ensure that they pay the correct amount of taxes but not more than required. The reason is simply that if an operation pays a relatively higher tax rate than its competitors in the same sector, it can be expected that the comparatively lower return on investment will make it a less attractive investment for potential providers of equity capital.

E. Lilford, P. Guj, *Mining Taxation*, Modern Approaches in Solid Earth Sciences 18,
https://doi.org/10.1007/978-3-030-49821-4_3

As discussed in more detail below, many components of expenditure can typically be offset against mining revenue, reducing the amount of taxable income that is available on which CIT will be levied. These offsets or tax deductions are of two types:

- those afforded to the bulk of corporate taxpayers irrespective of the nature of their activities (e.g. normal asset depreciation, interest expenses, losses carry forward etc.); and
- those relating specifically to the mining industry.

The latter are generally more generous than the economy-wide fiscal provisions (e.g. immediate or accelerated depreciation of exploration and development expenditure for mining specific assets, reduced royalty and/or CIT rates, tax holidays etc.) and generally represent a form of fiscal incentive to investment in the sector. Chapter 5 considers the justification for such incentives in recognition of a general perception that mining is conducted under a set of special conditions not normally encountered in other areas of the economy. It also considers the potential pitfalls of poorly conceived and/or understood fiscal incentives. Irrespective, the tendency in recent reviews of the fiscal regimes of various jurisdictions has been for increased harmonization of fiscal conditions throughout the various sectors of the economy.

On account of the greater opportunities for tax minimization in the context of CIT, the tendency in recent times has also been for governments to increasingly rely on the generally harder-to-avoid mineral royalty components of their fiscal packages.

In addition to royalties and CIT, government revenue is also generated by a large number of other taxes and imposts that, while individually less onerous, collectively represent a significant proportion of the overall government take. The majority of these imposts such as capital gain tax (CGT), VAT/GST, import-export duties and excises, withholding tax on remittance of interest and dividends abroad, stamp duties on transfer of assets, payroll and other financial transaction taxes, are applicable across the economy, but the impact of some of them may be more significant in the context of the mining industry. Others imposts, such as mining license fees and tenement rentals, various inspection fees, certain forms of shire rates etc. are by contrast industry specific. In addition most of these imposts tend to be hard to avoid and it would be inaccurate to make a sweeping generalization that resources companies may "structure away" their tax obligations as tax minimization opportunities tend to be primarily limited to their CIT liabilities which is only and not necessarily the major part of the government take.

Apart from the direct benefits to the State and to the operation's stakeholders that arise from the development and ongoing operations of a mine, its development also leads to the establishment of a secondary industry servicing the operation. Mining operations typically retain certain core activities, such as mineral processing, administration and geological (production) exploration, in house, utilising their own assets and their employees, while contracting other activities out to third parties better equipped and qualified to provide them at a competitively lower cost. These support activities can include:

- contract mining;
- maintenance contractors;
- support (rockfall prevention and stability, backfill, etc.) activities;
- rock mechanics monitoring and advice;
- sweeping and vamping contractors;
- environmental consultants;
- tailings management contractors; and
- step-out exploration drilling, amongst others.

Asides from the clear benefits to the operation, Government also receives additional Income in the form of taxes from these sources related to the mining industry, that is to say other than from direct royalties, CIT and other payments received from individual mining companies. These government revenues include personal income tax payable by all the employees of both the mining companies and of unrelated companies providing services to them and from the CIT and other taxes paid by the latter. The mining industry, while not a very large direct employer, generates significant economic multipliers whereby, according to the US Bureau of Economics (in Bivens 2019), for each direct job in mining, 3.9 jobs are created in other sectors of the economy through forward or downstream linkages, backwards or upstream linkages and other demand and fiscal linkages. Multipliers range between 2.3 for metal and non-metallic mining excluding fuel and 4.4 for coal mining and can be as high as 6.6 for the oil and gas industry. These multipliers represent very significant sources of additional government revenue indirectly attributable to the mining industry which should, together with any clearly identifiable positive externality, be considered by government during any policy formulation process concerning the industry in general and in the context of granting individual project approvals.

In the final analysis, for the development of mineral projects to occur, the host country must have the necessary political and economic attributes required to attract foreign investment, since much of the exploration and development capital will be sourced from other countries, as will some of the skills and services required during and after development.

Some of the development capital required will typically be sourced from equity markets around the world while the rest of the required capital may be secured as debt in the form of project finance from a financial institution or a syndicate of financial institutions. Either way, availability of the requisite development funds will be dependent on the recipient country displaying an appropriate level of political and economic stability and legislative attractiveness.

Of equal importance is the ability to repatriate funds from the project country so that adequate returns can be provided to the non-resident equity investors and also so that repayments of debt to the often non-local debt providers or lenders can occur.

As discussed in detail in Chap. 7, in addition to securing income to the state through the taxation package, a number of governments participate directly in the mining operation through the holding of a free carried or participative equity interest. Many countries also have policies that stipulate the involvement of local companies and / or local communities justified on the basis of the 'national asset' argument.

Their participation is often both equity and/or socially based, with their equity participation generally not being free carried.

Finally, although not discussed in depth in this book, the government of the day should be responsible for ensuring that the income it receives is wisely spent. It must be recognised and accepted that a mining operation and its associated minerals extraction and sales is a finite non-renewable activity and therefore that the income derived by the State from mining operations should be spent with that in mind. This does not mean that the State must necessarily put the money back into the mining industry, but it may undertake certain initiatives that, while providing a broader spread of benefits, may also assist in further resources development. As an example, developing additional power generating facilities with associated distribution and reticulation systems or, extending road and rail networks closer to or through mineral-rich regions, will, not only benefit the local communities, but also generate further exploration and improve the economic potential of possible future project developments in these areas. Types of common use infrastructure that can be funded by the State from resources and other sources of income can include:

- residential and industrial land developments and related road networks and other facilities close to potential mineral development areas;
- potable and industrial water supply (dams, bore fields, desalination plants, and related overland water pipelines, water storage facilities, etc.);
- landing strips and/or airports; and
- port developments for minerals and other commodities exports.

All of the above activities, processes, procedures and desires have an impact on a country's economic well-being and indirectly, therefor, will be caught up in some way or another in that country's taxation policies.

3.2 Mineral Royalties

A number of papers and publications have been drafted providing research and commentary around royalties in the resources sector as reviewed by Lilford (2017). The most authoritative and comprehensive publication among these is that supported by The World Bank and authored by Otto et al. (2006) titled 'Mining Royalties: Their impact on government, industry and civil society.' This publication details taxation in the minerals sector, reviews the mining royalty instruments adopted by various mineral-rich jurisdictions in the world, the impacts of royalties at the mining project level and on investors and society at large, as well as the governance challenge of effectively managing the benefits derived from royalties.

Four key types of royalties as put forward by Boadway and Keen (2014) typically apply to the minerals industry. These as discussed in greater individual detail in Chap. 4, include the:

- specific or unit-based royalty, where a specified amount is levied generally per tonne of ore mined;
- *ad valorem* or value-based royalty, where the royalty payment is derived by applying a percentage rate generally to the value realized on the first arm-length sale of a mineral product;
- profit-based royalty, where the royalty payment represents a percentage of the profit generated by the project as calculated according to the specific provisions of the relevant royalty regulations which may be at variance with those of the CIT legislation;
- economic rent based royalty, where the royalty payment is a percentage of the economic rent generated by the project computed on a cash basis after allowing a measure of 'normal' profit.

The specific/unit-based royalty and the *ad valorem*/value-based royalty are the royalty types most commonly imposed by governments on the mineral industry for the simple reason of producing relatively stable revenue flows and of being simpler to implement and manage and harder to avoid. Profit-based royalty are more rarely applied because of their less stable revenue generation and administrative complexity and, at the time of writing, economic-rent based royalties and 'product sharing' are limited to the petroleum industry.

Otto et al.'s (2006) publication provides the reader with a clearer understanding of the sensitivity of the performance of mining projects to royalties in quantitative terms making use of detailed financial models of three different typical mines hosting different commodities, i.e. copper, gold and bauxite. Their sensitivity analysis quantifies the degree to which the introduction of a royalty, if not economically efficient, adds to the unit cost of mining, thus distorting investment decisions and impacting the life of a mine by reducing the tonnage of ore that becomes economically extractable as the royalty rate is increased.

A limitation of Otto et al's work is that their worked examples and the related explanation simplistically reduce the lives of mines as the royalty expenses increase until negative cashflows are realized, without explicitly addressing the fact that their cut-off grade would vary as a function of changes in volume/weight-based or *ad valorem* royalty rates. The reality of these tonnage/grade trade-offs is more complex as will be detailed and clarified in Chap. 4 and in the Case Study outlined in Chap. 6.

3.3 Corporate Income Tax (CIT)

The globally tried-and-tested way of governments ensuring that they collect their due proceeds from profitable practices is through standard corporate income tax (CIT) measures. CIT will only provide income to the country's treasury if the operation in question generates an operating taxable income. Taxable income should, simplistically, be the residual amount remaining once operating costs and any other allowable deductions including duties, royalties and allowable capital recovery (e.g. depreciation and/or amortization), have been offset against a firm's revenue on an accrual basis. Corporate tax is then levied on the calculated taxable income generally at a fixed percentage rate.

One of the first factors reviewed by potential investors in the minerals and energy sector of a specific country is its CIT rate, which needs to be competitive with that of other jurisdictions to be attractive. Onerous CIT rates are a disincentive for investment because as this tax is levied directly against profits it reduces the firm's retained earnings, which represents the only distributable amount which can be used to declare and pay dividends to shareholders.

In the case of exploitation, not only does the project itself pay CIT on its profits to the State, but the individuals employed to work on the project pay personal income tax on their wages and salaries, which are also derived from production.

As an example, if it is assumed that the corporate tax rate is 30% and that 50% of operating costs relate to labour alone, with the average personal income tax rate being 30%, then it may be simplistically deduced that personal income taxes payable to the State arising from the operation amount to 15% of the operation's total income. If the operation is profitable, then an additional corporate tax of 30% of profits will also be payable, taking the effective overall project's tax rate to approximately 45% in this case.

There are, of course, additional taxes payable to the State beyond mineral royalties, corporate and personal income taxes, including taxes paid by shareholders on dividends, various forms of withholding taxes on the remittance of dividends, interest and other form of payments abroad, import and export duties and excises, land rates taxes, payroll taxes, stamp duties and many others minor taxes and imposts, collectively making the effective tax rate for most mining companies significantly higher. These various sources of income to the State arising from mining activities provide that economy with funds to benefit the nation's economy.

In essence, the rate of CIT in isolation may not be the main reason discouraging investment in the sector, when investors determine their potential return on investment (ROI), before making their investment decision often before having been granted security of tenure. Besides the various imposts listed above ROI will also be affected, by requirements for government carried and free-carried interests, policies around the expatriation of capital and prescribed local equity participation and procurement requirements. In addition, rules around limits on the use of expatriate skills may also influence an investment decision in the event that local skills and expertise is not readily available.

3.4 Capital Gains Tax

In a majority of countries disposal of mineral industry capital assets that result in a payment in excess of the written down value of the asset at the time of disposal, the so-called 'capital gain,' results in a tax being applied on the difference between the two. Different jurisdictions apply different specific capital gains tax (CGT) rates to the realised or deemed profit from the sale of capital asset irrespective of the time that has lapsed between the original asset purchase date and the effective date of the transaction or sale. A few tax regimes do not yet impose CGT, although an increasing number has been introducing it over recent times. The capital gains tax rate is often lower than the corresponding CIT rate, but the trend is once again for its progressive increase.

If by contrast, a capital loss has been incurred, the loss can typically be carried forward to be offset against future capital gains, but not against recurrent operating taxable income. In some but by no means most jurisdictions, capital losses may be offset against capital gains realized by an associated company.

3.5 Withholding Taxes (WHT)

A withholding tax (WHT) is typically a tax paid upfront to or withheld on behalf of the government by individuals and/or corporate organizations. The amount withheld is then credited against normal income taxes once final settlement of taxes is due.

In the context of foreign mining companies operating in an offshore jurisdiction, a withholding tax is an additional tax levied by the government on excess cash generated from operating activities, after all costs, taxes, royalties and duties have been deducted or paid, at the point where it may be remitted to foreign resident owners/beneficiaries. As withholding taxes are tantamount to being a tax on dividends they, not surprisingly, have a significant impact on determining the potential investment returns to, and proffered by, capital providers.

Overseas cash remittances on which WHT may be applied are generally derived from:

- dividends or interest payments to non-resident equity owners of mining projects in the country or to the providers of debt finance for their development and operations; and/or
- remuneration for the provision of goods and services by expatriates not being part of a permanent establishment (PE) in the country hosting the mining operations.

As a tax on dividends and interest, the two primary reasons why withholding taxes may be imposed by governments are:

- to restrict the amount of capital being removed from a country; and
- to garner or secure, not to be confused with "attract," additional income to the government from offshore investors. This is particularly the case, as discussed in more detail later, where there is evidence of significant shifting of profits abroad through multinational enterprises (MNEs) adopting contrived corporate structures designed to minimize their tax liability at the consolidated level.

While these reasons may be justifiable from the point of view of government, they nonetheless add to the burden of taxes and thus represent notable disincentives to foreign investment.

Table 3.1 Impact of 10% and 20% withholding taxes on effective royalty and tax rates

Comparative	Units	Assumed inputs	Effective rate
Gold price	US$/oz	1300.00	
Deductions including:	US$/oz	1080.00	
Operating costs	US$/oz	1000.00	
Depreciation allowance	US$/oz	50.00	
Duties	US$/oz	20.00	
Working capital	US$/oz	10.00	
Profit before tax	US$/oz	220.00	
Tax (30%)	US$/oz	66.00	
Profit after tax	US$/oz	204.00	
Withholding tax:			
10%	US$/oz	20.40	
20%	US$/oz	40.80	
Royalty proxy			
10%		1.57%	4.57%
20%		3.14%	6.14%
Corporate tax proxy			
10%		9.27%	39.27%
20%		18.55%	48.55%

Withholding taxes are only levied against earnings that are generated within the country and are to be expatriated. From the government's perspective, it could be assumed that the withholding tax may have the objective of:

- either merely increasing the revenues accruing to the government, or of
- encouraging the project owners to reinvest excess earnings back into the local economy rather than take that cash out of the country as a dividend or part-return on their investment.

The main issue with the second point is that logically, at some point, after reinvesting of retained earnings has been optimized and/or local attractive investment opportunities exhausted, the foreign project owners will eventually have no alternatives to taking excess cash out of the country where their next best investment opportunity is to be found.

In general, withholding taxes range between 10 and 20%. While this in isolation does not seem to be a particularly high range, it is additional to the official corporate tax rate already imposed. In effect a withholding tax could be viewed as an additional profit-based royalty but it would be differing from it in the respect of not being tax-deductible, that is to say applied below the line royalty. Depending on a number of factors including thc corporate tax rate, other payable duties, the total operating costs and the amount of depreciation expenditure incurred, a 10 to 20% withholding tax may equate to a royalty rate of between 1.5 and 3.2%, as shown in in the example of Table 3.1 below for a simplified gold project on a per ounce of sale basis. Typical royalties compare at between 2 and 7% worldwide, so if an operation is generating free, distributable cash flow and is already in a jurisdiction paying, say, a 3% royalty, the withholding tax may have the effect of significantly increasing the effective royalty payment. Again, this is a deterrent to investing.

Taking the withholding tax discussion to an effective tax on profits (normal corporate tax), the same 10 to 20% withholding tax range may be viewed as increasing the corporate tax rate by between 9 and 19%, as shown in Table 3.1 below.

3.6 Value-Added, Import-Export Taxes and Excises

It is estimated that value added tax (VAT) or its equivalent good and services tax (GST), is used in more than 160 countries worldwide. VAT/GST is added to the sales price of a product whenever value is added to that product at each stage of its supply chain. It is a consumptive tax, simply being the tax applied to the purchase of a good or service, and is based on the cost of the product less any costs attributed to the material costs incurred in making the product to which value has been added that have already been taxed. Advocates of VAT state that it raises government revenues but fails to punish wealth, while critics state that it is a regressive tax, placing an increased economic pressure on lower-income residents. Either way, VAT is levied on the production and sales process at each point where value is added and a sale occurs.

An exemption from VAT on mining equipment is a fiscal incentive in many jurisdictions. Unfortunately, as will be discussed in more detail later in Chap. 9, the government process of collecting VAT upfront and later refunding it to companies appears to create significant administrative difficulties in many developing countries.

Many jurisdictions levy imposts on the importation and exportation of goods in general. In the context of mining products import and export taxes are typically charged against the importation of crude ore or partially-beneficiated mineral products, such as including mineral ores, concentrates, smelter mattes and other unrefined metals and minerals. These taxes are in many respects similar to the import/export tariffs and duties levied against completely processed, final products such as refined metals, equipment and other tangible goods.

For example, Zambia introduced a 5% import tax in 2019 on ore and concentrates imported for further processing in that country. The reason put forward for introducing this import tax was to incentivize the owners of domestic processing plants and associated facilities to source their primary feed for their Zambian plants within the country's borders. This in turn would purportedly increase employment within the country, providing significant benefits to the local population as well as providing additional income through various taxes levied by the State.

Export taxes follow a similar pattern to import taxes, other than the charge is levied against under-processed materials and unfinished products that are sold into foreign markets for further processing, refining and fabrication. The thinking behind imposing export taxes on crude or partially-beneficiated minerals is to provide an economic incentive to the minerals producers to establish in-country processing facilities in order for the domestic economy to capture more of the value-adding benefits arising from further beneficiation and fabrication.

Excise taxes may either be levied as an ad valorem excise or as a specific excise. An ad valorem excise tax is charged at a fixed percentage of the revenue or income received by the seller of the goods or services. Therefore, as the price for the goods or services changes, the amount paid in ad valorem excise taxes also changes. If the tax is a specific excise tax, then it is payable as a fixed dollar amount per unit sold. In general, excise taxes are applied to goods that are perceived to reflect a notable social cost, including to fuel on account of pollution and road tear and wear considerations and alcohol and cigarettes under the health point of view. In these latter case, the excise tax is often referred to as a sins tax.

3.7 Quasi-Taxes

In general, a quasi-tax refers to the monetary value of a contribution to the development of local infrastructure as well as other community-beneficial infrastructure and services within a country or region affected by a mining operation. Depending on their nature some of these expenses may or may not be able to be offset against the taxable income of a mining company for the calculation of its corporate income tax. In the former case they represent a legitimate cost of carrying out the mining business in the latter case an additional tax.

The imposition of quasi-taxes may apply to a mining development/operation either under the general requirements of the prevailing mining legislation or under the specific terms and conditions attaching to the mining title or contract for individual projects. As such quasi-taxes, by contrast with mineral royalties and corporate tax, do not directly contribute to government's revenue and are typically not stated in terms of a rate, but they are quantifiable in terms of the contributions they make to the satisfaction of specific community or infrastructural needs. They nonetheless may indirectly affect government revenue by removing potential sources of expenditure when a company's contribution to common use physical and social infrastructure exceeds the extent of the direct impact of a project on the communities it affects. By way of example, a development or redevelopment tax, education tax or skills tax, often individually or collectively referred to as a skills development levy, are all quasi-taxes and typically range between 2% and 5% of taxable income (South African Revenue Service, sars.gov.za; Ramdoo 2018).

The skills development tax or levy is an additional tax charged against taxable income that is supposedly used solely for the training, upskilling, development or reskilling of the employees of an operation. The overall intent of this tax is for the benefit of employees who may, at some point, leave their current employ and find employment elsewhere. It may sound counter-intuitive to provide training to group of employees with the knowledge that they may eventually leave the company, but when one considers that mining is a finite activity and that all ore bodies eventually will be depleted, the upskilled or retrained employees will be better-placed to find alternative employment when operations cease. While the levy may be considered as being a contribution to socio-economic sustainability, this notion is dependent on how the collected taxes are spent. It is imperative that a skills or development levy, which is paid to the State, be used for the purposes for which it was intended, otherwise it just becomes another form of royalty, adding to the cumulative burden of the mining tax package.

The development and operation of mineral projects over its value-adding cycle from mineral exploration to development and exploitation of a mine, to mineral processing, marketing and site rehabilitation leads to significant demand for a variety of both basic and specialized technical and administrative skills in the field of mining and associated industries. As already indicated, in the case of developing countries, these skills may be scarce or even unavailable and initially tend to be provided by expatriate employees, contractors and consultants. Most jurisdictions require progressive skills transfer and development which may be codified in the host country's regulatory and taxation regime in the form of a specific skill-levy and/or community training obligation as conditions of the mining title.

Talking about exploration first, it is a generalisation to say that it requires the same or at least similar skills. While this may hold true in many cases, there are a range of minerals and associated geological conditions that do not lend themselves to standard exploration techniques. For example exploration for potash, a source of potassium used in fertilisers, involves specific drilling conditions as most potash occurrences are associated with salt, which would dissolve if water were used as the drilling medium. Instead, a saturated mud must be used, and then only after lining the initial drill-hole with a cemented steel drill column. As a consequence, the skills necessary to drill for potash are very specific, and closer to the skills required for the drilling of exploration and production wells in the oil and gas sector.

Moving from exploration to mining, different orebody configurations, geotechnical ground conditions and different minerals lend themselves to different mining methodologies. Even the same mineral commodity may lend itself to a variety of mining methods. For instance, coal mining operations may be carried out in open cut or underground, with the latter using a board and pillar mining method either with continuous miners and shuttle cars or using drill and blast methods with a coal cutter, face scoop and shuttle cars, or as an alternative the coal deposit may be mined on a longwall or shortwall basis.

Similarly hard rock operations to extract ferrous, base and precious metal, non-metallic industrial minerals, diamond etc. also have numerous and diverse mining methods available to them. These range from large open cut mines paired by extensive transport systems, to large underground block-caving operations, to more selective sub-level caving to simple supported and unsupported stoping mining methods, all of which requires different skills.

The socially-based participation of the local community or associated companies seeks to provide skills and training that not only can be employed at the mining operations but that find application in the wider mining and non-mining communities. Typically appropriate training is offered on site to members of the local community, as well as in the form of bursary or scholarship schemes to assist current and potential employees in further educational development in off-site training institutions. Where necessary and desired, many communities are also offered English literacy classes to bring their reading, writing and oral English skills to higher levels.

Post-secondary educational opportunities within the resources sector are generally abundant in economies that are supported by an active mining industry. Consequently, higher (tertiary) education is also required in the field of mining and associated industries to support the development and continuing operations of a mine. Beyond just the undergraduate sector, trades and post-graduate education are also critical to feed into the sector. While the various trades and graduates necessary to perform various functions on and with the mine are critical, it is wise to acknowledge and support post graduate research in the industry to improve existing processes and to innovate new processes.

Arising from employees' turnover and after the closure of a mining project, the people who have been trained are likely to continue to hold jobs. Through these social requirements, the government continues to benefit from taxes resulting from both their direct employment at the operation as well as future employment as those same skills are utilised elsewhere in the country in either the mining or in a non-mining industry.

Take Away Points

Government's point of view

- In general the bulk of mining taxation revenue should be derived from mineral royalties, mostly specific and ad valorem, and corporate income tax as the main instruments of a mining taxation package. Royalties provide revenue stability, ease of administration and are hard to avoid, while CIT is more economically efficient and equitable but produces uneven revenue flows, is more easy to avoid and harder to administer.
- The specific characteristics of the mining industry warrant the application of some fiscal incentives preferably in the form of immediate deductibility of exploration expenditure and accelerated depreciation of mine development and construction costs.
- Tax holidays are generally less desirable, but if required, need to ideally be limited to CIT and be structured with extreme care.
- There is merit in applying favourable taxing measures to gains realised on disposal on exploration tenements as this constitutes a significant incentive for junior and mid-sized entrepreneurial exploration companies to invest in the country.
- Review the degree to which existing and proposed double taxation agreements would have the effect of reducing withholding taxes to a point where it would offset all other potential benefits the country may derive from them.
- There is some justification for providing exemption from custom duties of mining plant and equipment and from VAT for export orientated companies.

- Government should attempt to cut the red tape and limit the number of nuisance imposts that raise little revenue and complicate approval and administration processes.

Mining industry's point of view

- There is no point in comparing the mineral royalty or the CIT rates of different jurisdictions in isolation. The choice of an investment destination must be based on a comparison of how different countries' tax packages impact on the profitability of specific projects.
- In general weight and/or value based mineral royalties will be more clearly understood by both parties and easy to administer than profit-based ones, thus reducing the incidence of disputes.
- While tax holidays may be very desirable, they need to be structured and understood clearly by both parties if they are not to result in later disputes. This may involve, asides from timing, constraint as to the tonnages of ore of contained metal that may be extracted over the holiday period.
- Being cognisant of the impact on withholding taxes of the current changes to tax treaties under Actions 6 and 15 of the OECD's BEPS initiative.
- Establish clear arrangements/protocols for the refund of VAT and other duties thus avoiding the significant delays experienced in many jurisdictions.

References

Bivens J (2019) Updated employment multipliers for the US economy. Economics Policy Institute, Washington, DC

Boadway R, Keen M (2014) Rent taxes and royalties in designing fiscal regimes for non-renewable resources. CESifo Working Paper No 4568. Category 1, Public Finance

Lilford EV (2017) Quantitative impacts of royalties on mineral projects. Resour Policy 53 (2017):369–377

Otto J, Andrews C, Cawood F, Doggett M, Guj P, Stermole F, Stermole J, Tilton J (2006) Mining royalties – a global study of their impact on investors, Government and civil society. The World Bank, Washington, DC

Ramdoo I (2018) 8th global commodities forum. Skills development in the mining sector: making more strategic use of local content strategies. United Nations Conference on Trade and Development, 23-24 April 2018, Geneva

South African Revenue Service. https://www.sars.gov.za/TaxTypes/SDL.aspx

• ... should attempt to curtail red tape and limit the number of instances of processes that cause little revenue and complicate approval and administration processes.

Mining industry's point of view

• There is no point in comparing the marginal royalty or the CIT rates of different jurisdictions in isolation. The choice of an investment destination must be based on a comparison of how different countries ... in ...

Chapter 4
Different Types of Mineral Royalties

Abstract Mineral royalties are generally viewed as a compensation to the state for the acquisition of its non-renewable resources. Royalties may be applied on different bases, either volume, value at various points along the value chain or profit/rent, in different jurisdictions. While there is no correct or incorrect version of a royalty, each different royalty type will achieve different and at times incompatible government objectives and will have a different impact on both government revenue versus the income of mining companies and consequently influence their willingness to invest and the mode of mining development, operations and financial performance of projects.

Keywords Mineral royalties · Royalty rates · Royalty types

4.1 General Principles

Most research to date delves on providing insight, discussion and guidance around taxes and royalties in general (e.g. Hartman and Guj (2013) and on quantifying income levels and State benefits (Smith 2013), but does not discuss to any great extent the direct and indirect impacts that the imposition of mineral royalties may have on the resources and financial performance of mining companies and on creating a threshold for further investment. Quantification of the impacts on orebodies and resource companies arising from royalties, can be found in Hotelling (1931), Smith (2013) and above all in seminal research on the nature of mineral royalties world-wide and their impact on government, industry and civil society is provided by Otto et al. (2006). Finally Guj et al. (2013) focus on their administration as a main component of the mining fiscal regime package of developing countries.

More specifically, Smith (2013) provided an overview of economists' understanding of the impacts that taxes have on the extractive resources sector. His work focused on research methods and techniques and not on the quantification of the impacts. He took his discussion into an overview of the various approaches that lie between reservoir simulation models and the neoclassical theory of production to highlight his tax policy analysis. Smith (2013) lists numerous studies that detail taxes

E. Lilford, P. Guj, *Mining Taxation*, Modern Approaches in Solid Earth Sciences 18,
https://doi.org/10.1007/978-3-030-49821-4_4

and royalties, with as many as ten of the studies focusing on the non-renewable resources sector. However, the studies focussed on the appropriateness, value and State benefits arising from these sources of income and minimally considered the implicit impacts on resources sterilization, less-than-optimal extraction and other operator impacts. Most of the remaining studies focus on oil and gas.

Philosophically industry practitioners and academics are divided as to what a mineral royalty really constitutes. The prevailing view is that a mineral royalty represents the consideration paid by a mining company to the state for the right to mine and sell the latter's non-renewable resources. Accordingly royalty payments should be a function of the amount or value of the resource as taken from the ground, i.e. of broken ore or petroleum at the point of extraction be either at the mine head or well head. In other words, as under this definition a mineral royalty is the purchase price for the extracted resource, strictly speaking it is not a tax. As a consequence, no royalty should be levied on the value added to mineral products downstream the value chain from the mine head. Proponents of this philosophy, which is favoured by most mining jurisdictions, the mining industry, many academics and the World Bank (WB), consider production-based royalties using a measure of either the physical volume/weight or the value of the mineral extracted or sold as the most appropriate base on which to apply a royalty rate.

The alternative view, supported by many in the petroleum industry, many pure economists and the International Monetary Fund (IMF), considers a mineral royalty as just another mineral taxation instrument to appropriate a share of the profit or economic rent generated by a mining project. Proponents of this philosophy consider royalties based on profit or on economic rent more desirable.

Whether mineral royalties are viewed as taxes or not may have significant legal and political implications. This is because their introduction may be hindered under the constitution of many countries which generally mandate that any new tax should be subject to a popular referendum. The recent attempt by Chile to introduce a royalty on copper production, for instance, was initially frustrated because considered unconstitutional.

In many federations, such as for instance Australia where the powers of the central federal government are specified in the constitution, any other powers on which the constitution is silent, such as land and resources management, fall in the ambit of state or provincial legislature. In the overwhelming majority of instances mineral royalties are not viewed as a tax but as a cost of carrying out the business of mining and therefore represent a legitimate deduction for the purpose of determining the taxable income on which income tax is then levied by the federal government.

As already pointed out, determination of a royalty payment entails the application of a royalty rate to a base. The rate can either be a monetary quantum or a percentage. While the base can be either a physical (e.g. volume or weight of production) or monetary (e.g. value of production/revenue or profit/rent). Consequently and excluding hybrid structures, and as detailed in Boadway and Keen (2014), royalty types include:

- specific or unit-based (on weight or volume);

- value based or ad valorem (on sales revenue);
- profit based;
- economic rent based; and
- production sharing contracts.

The first two royalty types listed above, and to a lesser degree the third listed type, are of common application to the mining industry, while the last two are more commonly encountered in the petroleum industry and currently only exceptionally in mining.

In addition, there are a number of hybrid royalty formulations, primarily profit-based royalties subject to a minimum, generally a specified amount or more often a value-based charge.

The term 'quasi royalties' is often used with reference to other forms of impost common in developing countries such as the requirement for non-contributory or free government equity in projects, which will be discussed in detail in Chap. 7, and others such as contributions to common use infrastructure and facilities under the terms of mining contracts.

As already discussed, determination of an appropriate royalty rate is a matter of government policy as to how much and what proportion of the economic rent it wishes to appropriate using a royalty as a levying instrument and where it wishes to position the country in terms of its international competitiveness for foreign direct investment (FDI) in the fiscal context. Boadway and Keen (2014) discuss issues surrounding the determination of an optimal royalty rate for resources projects.

Determination of an appropriate and reliable royalty base, by contrast, should be a matter of facts. This can be simple for specific royalties where the physical quantities on which they are based can be measured directly relatively easily by means of weight bridges and such a like.

As it will be discussed in detail below, the process becomes increasingly more complex for value-based royalties because very few mineral sales involve broken ore at the mine head or even at the mine gate and the corresponding royalty base values must be derived ideally from the prices realised for the first mineral products sold at arm's-length after some value has been added by processing downstream from the mine head along the value chain. Sales may involve a range of intermediate products of various specifications sold in opaque markets, all the way down to sales of refined metals in transparent and regularly priced terminal markets.

Although there are some differences in the way the profit/rent base normally used for royalty purposes is calculated, the approach is not dissimilar to that adopted to determine the taxable income for corporate income tax (CIT) and, while this type of royalty system is economically efficient, it presents all the accounting complexity and opportunities for avoidance, particularly in terms of allowable deductions, inherent in the federal income tax. Given this similarity, one may also justifiably question the administrative wisdom of combining these two essentially similar instruments in the same fiscal package.

The following sections highlight the characteristics of different royalty types in more detail, the degree to which they achieve the various government's objectives as outlined in Chap. 2, and compare the sensitivity of the related payments to changes in commodity prices.

Some companies hedge a percentage of their production in the forward markets to reduce their exposure to the risk created by the commodity price volatility. As a consequence the revenue realized on the sales of their mineral products may be different from what it would have been had they sold their metal on the spot market. It can be argued that the value of the mineral at the time of sale, for the purpose of levying a value-based mineral royalty but not a profit-based tax, should be based on the prevailing spot price since hedging gains or losses are the result of a business decision which has nothing to do with the value of the mineral in the ground or at the minehead. As a consequence in most jurisdiction an *ad valorem* royalty is calculated as a percentage of the commodity sold multiplied by the spot commodity price at the time of the sale transaction, not the forward price actually realized. This fact may work either in favour of, or against government's revenue, depending on whether on the day of transaction the forward price happened to be lower or higher than the corresponding spot price. However, in the former case, because the operation's income is negatively impacted by a below-spot-price hedge, there is a risk that the resulting hedging loss may precipitate the company ceasing operations to the detriment of both the company's shareholders and government. In reality, to the extent that metal market are most of the time in contango, that is to say in a situation where forward prices are higher than the corresponding spot prices, hedging has tended to favour the performance of those company that actually have the capacity to deliver the physical metal at the expiry date of their forward sales, rather than just being market speculators.

Similarly to corporate income tax, if the royalty is profit based, the regulators will simply look at the sales profit realized by the mining operation irrespective of whether it sold its product forward or spot, thus taxing any resulting hedging profit but not necessarily providing relief for possible hedging losses.

Chapter 6 will provide a model-based, realistic case study of how different levels of royalty rates and other imposts on a gold mining project affect, not only the sharing of the economic rent between government and industry, but also the degree to which resources are sterilized leading to their suboptimal economic exploitation. As already discussed in Chap. 2, sterilization occurs because, in an endeavor to maintain profit margins, mining companies will tend to react to increases in inefficient taxes by lifting the cut-off grade, that is to say by preferentially mining higher grade material leaving behind lower-grade resources. In the process mineable reserves are reduced and the life of the mine (LoM) shortened relative to what it would have been at a lower level of imposts.

Contrary to Burness (1976), who argued that the life of an operating mine remains unaffected by a royalty, Gillis et al. (1978) identify a number of mining operations in Bolivia that were prematurely closed due to high-grading arising from the introduction of a royalty. Further evidence of the economic impact of high-grading on a mining project is discussed in Conrad and Hool (1981), Lockner (1965), Steele (1967), Laing (1976), Gillis et al. (1978) and Conrad (1978) and Lilford (2017).

4.2 Specific, Volume or Weight-Based Royalties

A specific royalty applies a fixed amount per unit of volume or weight of ore produced or sold, e.g. $\$/m^3$ or \$/t. Under some jurisdictions the rate is subject to regular or occasional indexation.

Specific royalties are economically inefficient in so far that payments constitute a cost of doing business which is unrelated to either sales revenue or profit.

They are, however transparent and easy to explain/understand as they are closely related to the mining process.

They are also simple to administer, relatively easy to audit and hard to avoid.

Specific royalties provide predictable and stable revenue as long as a mine operates, but may precipitate mine closure if the price of the mineral product falls. However, not being related to the price of the commodity, collections remain the same also during commodity price boom periods, which may lead to public dissatisfaction and a perception of not capturing a 'fair share' of profit and political pressure for government to increase the royalty rate.

In general specific royalties are widely applied to low-value, bulk materials (e.g. sand, gravel aggregate etc.) and in some jurisdictions to higher-value bulk mineral (e.g. coal, bauxite etc.).

4.3 Value-Based Royalties

In a value-base or *ad valorem* royalty a percentage rate (typically 2–10%, but mostly 2.5–5%) is applied to the realised or estimated sales value of mineral products.

Ad valorem royalties are more economically efficient than specific ones as, being a direct function of commodity prices, payments increase with rising prices and decrease with falling ones softening their impact on the miners. As a consequence, payments will be variable but royalty will still be payable even if, under conditions of falling prices, a mining project may become unprofitable as long as it continues to operate.

As discussed in more detail below, depending on the nature of the mineral product sold and its marketing characteristics there may be a choice of possible taxing points and related royalty value bases. Accordingly *ad valorem* royalties can be simple to relatively complex to administer. Increasing administrative complexity increases compliance costs as well as opportunities for minimisation or even avoidance.

Value-based royalties are by far the most common type of royalty applied worldwide to base and precious metals and other high-value mineral commodities.

4.3.1 Determining the Value Base at Various Taxing Points Along the Value Chain

Very few sales of broken ore take place at the mine head, but most mineral products are sold after some value has been added to the crude ore by downstream processing along its value chain, through mineral concentrates and other intermediary products, all the way down to crude and refined metals.

Transparency and administrative efficiency dictate that the royalty value base be derived with reference to the price realised in the arms-length sale of the first mineral product sold along the value chain. As shown in Fig. 4.1 the point at which a royalty value may be based can be set at different levels of processing along the downstream value chain such as:

- ex-mine (V1) at the mine head but more frequently at the mine gate/ore ROM stockpile after crushing and screening;
- FOB port of export (V2) after trucking and or railing and ship loading;
- at the smelter (V3) for concentrates and other intermediary products after sea freight (CFR basis) or freight and insurance (CIF basis); or
- in terminal (V4) for refined metals after smelting and refining.

In effect the ex-mine value will be a function of the price of the commodity, the grade and quality of the ore body, the cost of processing the ore into a sellable mineral product and distance from markets.

In line with the principle that a royalty should not apply to value added downstream from the mine head, to achieve a consistent payment relative to the mine head value, decreasing royalty rates should apply to the price realised the further downstream a mineral product is sold.

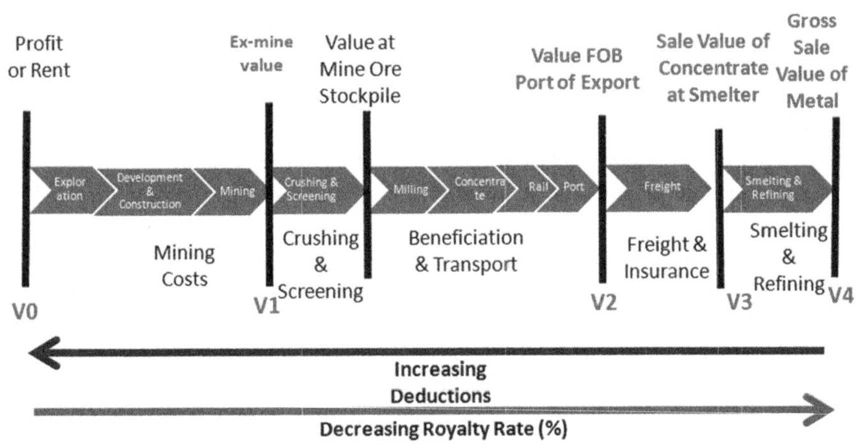

Fig. 4.1 Decreasing *ad valorem* royalty rates with increasing beneficiation

In spite of this, some jurisdictions apply a single royalty rate to a mineral commodity irrespective of it being sold as different mineral products. This approach:

- under-taxes mineral products to which little value has been added; and
- over-taxes more highly processed products downstream the value-adding chain; and
- encourages high-grading and discourages investment in downstream processing.

There are two approaches that can be followed to obviate this problem, namely to either apply a:

- decreasing royalty rate as the taxing point moves progressively downstream; or
- single royalty rate to the ex-mine value of the mineral obtained by netting back from the price realised on the first arm-length product sale the cost of all processes incurred downstream from the mine head.

In line with the first approach, some jurisdictions, such as Western Australia, apply discrete standardised royalty rates decreasing from 7.5% for crushed and screened ore, to 5% for concentrate and to 2.5% for refined metal sales.

While administratively simple, this approach appears to achieve to some degree the government objective of levying royalties approximately equivalent to 10% of the ex-mine value, albeit on average and not for individual commodities, thus improving economic efficiency and encouraging investment in downstream processing.

The Western Australian system, however, has its own drawbacks as it implies that:

- smelting and refining costs represent 50% of the value of the refined metal;
- the cost of concentrating and transporting mineral products represents 33% of the value of the refined metal; and
- crushing and screening the ore represents 25% of the value of the refined metal.

This is assumed to be the case for all commodities and all mines in WA, which is clearly not the case. High-grade mines are therefore undertaxed, while lower grade mines with complex processing systems are comparatively overtaxed. These systemic weaknesses were clearly identified and discussed in a review of the Western Australian mineral royalty system conducted by Bradley.

In its recent 'Department of State Development and Department of Mines and Petroleum (of Western Australia) 2015' (Final Report 2015) the Western Australian government recommended an increase in the royalty rate for gold from the current 2.500–3.725% on the basis that its processing and marketing costs are relatively lower than those incurred for most other metals. The gold industry mounted a very successful political campaign to oppose the change, which while still on the government's agenda, to date has not yet been implemented.

In addition to the extent that royalty on most mineral products is applied on their value FOB port of export, domestic transport and port loading costs are not deductible and as a consequence mines further inland pay proportionally higher royalties. As a result, while the Western Australian royalty system is administratively simple, it goes only part of the way towards improving economic efficiency and equity in that it inhibits development of marginal ore in remote locations and does not recognise the different ability to pay of various projects. By contrast, as an incentive for investment in downstream processing in the country, domestic transport costs are deductible if the ore/concentrate is to be conveyed to a processing plant located in Australia.

The current Western Australian royalty system may also represent a disincentive to invest in downstream processing and create inequity and administrative difficulties if the product sold is neither ore, concentrate or metal, but an intermediate product, e.g. vanadium pentoxide, titanium dioxide etc. This type of difficulty has recently been recognised by the Western Australian Government in the case of royalties to be applied to magnetite iron ore concentrates on account of its production costs being higher than that incurred in the simpler process of crushing, screening and minor blending of direct shipment ore (DSO).

4.3.2 Taxing Points Characteristics

Theoretically V1 should be located at the point of extraction or mine head. In practice no sales of broken ore take place at this point. As a result, royalties tend to be applied at the so-called mine gate, that is to say when sales are from the run of mine (ROM) stockpile, after crushing, screening and, in some cases blending, has taken place. However, commercial sales at this point are also rather rare and often carried out between associated parties, hence in the absence of an active independent market reliable pricing information is scarce.

In practice, in the vast majority of cases in which a royalty rate is applied to the mine gate or ex-mine value of sales V1, this value has been derived by netting back from the price realised in the first sale of a downstream derived mineral product to a third unrelated party all the processing costs incurred upstream of this point of sale back to the mine gate. As discussed later on, the mine gate value was selected as the best taxing point to use in the determination of the taxable income on which to levy the, now rescinded, mineral resource rent tax (MRRT) on production of iron ore and coal in Australia.

Of course estimating the allowable deductions necessary to carry out the net back processes requires access to confidential company cost data, which companies may be reluctant to divulge. As it will be discussed in more detail in Chap. 9, the process becomes particularly difficult and opaque because of transfer pricing considerations if the first arm's-length sale occurs through the intermediary of a related marketing hub domiciled, as often is the case, in a foreign low-tax jurisdiction.

In spite of these administrative complexities, V1 is closer to the resource value, hence consistent with royalty principles and more economically efficient. The revenue generated by levying royalties at this point is closely linked to mining activity, more economically efficient and equitable to companies reducing their incentive to high-grade. From the point of view of government revenue tends to be less stable and predictable.

The value of sales FOB port of export V2 is a widely adopted base on which to levy mineral royalties. It is typically adopted for some base and precious metals concentrates and for bulk exports of iron ore, coal, bauxite, manganese etc.

As for V1 determination of this value base requires netting back from the price realised in the first arm's-length sale confidential company data relating to freight and a few other costs incurred in transporting the mineral product to the point of sale if the sale is on a CFR basis, or including insurance if the sale is on a CIF basis.

Royalties levied on this basis are reasonably transparent and easy to explain as they are closely linked to the mining process and generate reasonably predictable revenue. They also represent a middle-of-the-road compromise in terms of economic efficiency and equity, reducing but not altogether eliminating the incentive to high-grade, if a single royalty rate applies for the mineral commodity.

The value of sales at the smelter V3 realised in a smelter sale on a CIF basis, the so-called Net Smelter Value (NSV), is also a commonly adopted royalty base because only a minority of very large base metals mining operations are fully vertically integrated, with the majority of medium- to small-size producers, which cannot justify investing in integrated smelting and refining facilities, selling base metals (e.g. copper, lead, zinc, nickel) concentrates and associated gold and silver credits, to smelters.

Further upstream the Net Smelter Return (NSR) is obtained when all transport, insurance and other related costs incurred downstream from the concentrator, both domestic and international, are netted off the NSV.

Depending on the metals involved, smelter and refining contracts and their related terms and costs are relatively well established, benchmarked and consequently easily audited. Table 4.1 provides a numerical example of how to calculate the NSV (i.e. royalty base value V3), the FOB royalty base value V2 and NSR for the sale of a copper-gold concentrate.

As a NSV is derived well down the value-adding chain a royalty based on it is more efficient and equitable to companies if a single royalty rate applies thus lowering the incentive to high grade. In practice, as shown in Fig. 4.2, mineral royalties are often levied on the free on board (FOB) value V2 of the concentrate at the port of export.

The unit deduction and treatment and refining charges are higher when there is ample supply of Cu concentrates and/or Cu prices are high. Under these pricing circumstances the price participation component may also become particularly contentious because it was originally set at times when the normal price for copper was in the range between \$0.80 and \$0.90/lb. The idea was for miners to receive support in case of price falls and to share in their good fortunes with smelters in case of price rises. However, with significant rises in price in recent years price participation has become a significant component (up to 50%) of smelters charges and has become resented by the miners.

Credits for by-product metals, particularly precious ones, can be, as in the case of Table 4.1 a significant component of the price realised on the sale of concentrates.

Table 4.1 Calculation of the net smelter value

Net smelter value (NSV) of Cu-Au concentrate (US$/T)			
Formula = Net smelter value (NSV) = [(M – D) * (P – R)]/100 − T − PP + C			
Assumptions		**Calculation**	
M = Grade of concentrate (Cu %)	30.0%	GV = Gross value of Cu in concentrate (US$)	**1734**
G = Gold grade (g/t Au)	25	Less:	
D = Unit deduction Cu (%)	1.2%	Unit deduction charge	−69.4
PCu = Metal price (US$/t Cu)	5780	T = Treatment charge	−77.0
PAu = Gold price (US$/oz)	1515	RC = Refining charge	−44.4
RC = Refining charge (US$/lb. of metal)	0.07	PP = Price participation charge	−166.5
T = Treatment charge ($/t of concentrate)	77	Plus: C = Net gold credit	1169.0
PP = Price participation rate e.g. +10% of price above $0.9/lb. or − 10% of price below $0.80/lb	0.1	**V3 = Net smelter value at smelter (NSV) (US$)**	**2545.7**
C = Approximate gold credit (%)	96%	Less:	
CF = Conversion factor lb to Kg	0.4536	S = Sea Freight	−47
S = Sea frelght, loading and insurance (US$/t)	47	**V2 = FOB value of concentrate at port of export**	**2498.7**
R = FOB royalty rate (% of FOB value)	**5%**	**FOB royalty (US$)**	**124.9**
K = Distance mine to port (Km)	300	Less:	
TR = Railing/trucking cost (US$/tKm)	0.09	TRC = Railing/trucking cost	−27
		NSR = Net smelter return at mine (US$)	**2471.7**

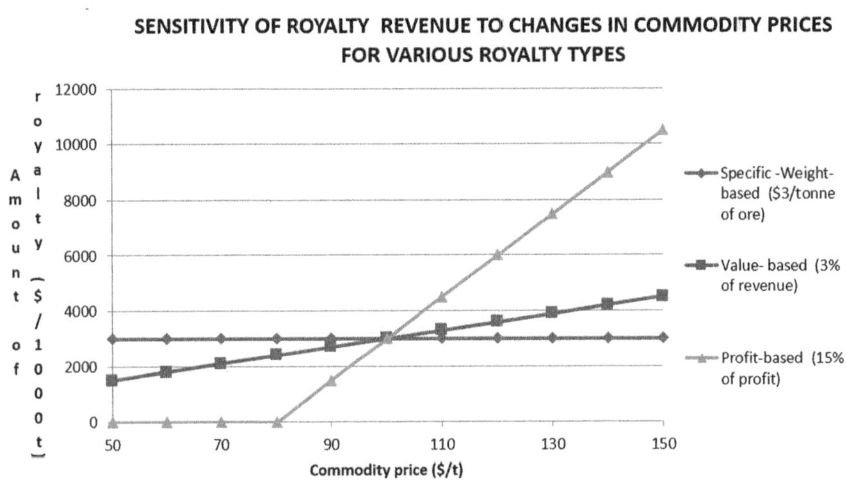

Fig. 4.2 Sensitivity of royalty payments to changing commodity prices based on royalty type

Conversely penalties for detrimental metal impurities can result in heavy discounts in prices. This is particularly the case for some 'dirty' lead-zinc concentrates, which at the limit may become unsellable. The contract of sales will specify the percentage of the market value of the by-products which will be payable and the discount to be applied for the degree by which impurities exceed the acceptable limits as also specified in the contract.

Using the price realised on the sales of refined metals V4 as a taxing point has some distinct advantages. These stem from the fact that most refined metals are sold in well established, transparent and active terminal markets (e.g. London Metal Exchange (LME) and London Bullion Market (LBM)) requiring very stringent product specifications. Of course where metals are sold in a crude unrefined form, e.g. gold-silver dore,' blister copper, nickel matte etc., the refining costs to make them compliant with LME specifications will need to be netted back. Differences may also arise on account of different warehousing considerations. A provisional price may be applied to determine initial and progress payments for individual shipments and subsequently adjusted on the basis of the price realized by the smelter on the sale of the refined metal(s) including related credits and penalties for associated precious and other metals.

The transparency of the market and the close functional link between revenue and mine activities make the overall process easy to explain to industry and other stakeholders. From the government's point of view the ready availability of relevant transaction prices from terminal markets makes administration and auditing of royalty returns and related invoices comparatively easy. In addition revenue collection tends to be relatively predictable and stable.

Use of V4 as a royalty base becomes inequitable to mining companies if a single royalty rate applies to all products of a given mineral, which, if high, encourages high-grading with concomitant sterilisation of some of the mine resources as discussed in detail in Chaps. 2 and 6 and constitutes a disincentive to investing in downstream processing value-adding facilities.

The V4 value base is typically used for gold, silver, PGMs, aluminium, base metals and nickel etc. sold in refined metallic form on the LME and LBM, which regularly publishes their daily prices.

Complication arises when, as in the case of many bulk minerals, sales are not on a spot basis but carried out under the terms of off-take agreements. These may specify that the price to be paid, as for instance in the case of iron ore, will be determined as the average over a specified period of time, e.g. 3 months, with reference to a regularly quoted spot price index, e.g. IONEX index. As indices are regularly published for iron ore fines with iron content of 62 or 58% and standard content of aluminium, silica, phosphor, sulphur etc. the applicable price must be adjusted for differences in iron grade and impurities resulting in price discounts relative to that quoted for the index.

Transfer pricing considerations may make the process administratively complex and opaque from the government's point of view, as discussed in more detail in Chap. 9, unless the relevant mining contracts contain specific obligations for the mining company to disclose for auditing purposes the actual invoices relating to mineral sales and transport costs irrespective of whether they were incurred by the company or an associated marketing hub company domiciled abroad.

4.4 Profit-Based Royalties

In profit-based royalties a percentage royalty rate is applied to a measure of "profit" estimated at project level. The royalty "profit" base is generally different from the taxable income on which corporate income tax (CIT) is levied primarily in the way capital recovery and financing costs are treated.

A profit-based royalty is economically efficient as it eliminates the incentive to high grade and equitable in that rich mines close to market with greater ability to pay will in fact pay a comparatively higher royalty.

From the government's point of view, however, profit-based royalties present a number of disadvantages. First among these is the variability in revenue, which may be very low or even absent in the early years of production and/or when commodity prices are low. They are also relatively complex to administer and audit and, being project specific, may generate cost allocation issues relating to assets which may be shared with other company projects. In addition being poorly matched to mine production activities they tend to be less transparent and therefore difficult to explain to stakeholders. In addition to the extent that the profit base on which successive royalty monthly or quarterly returns during a year are based is provisional, payments will need to be subject to adjustments after the annual accounts are finalized and audited.

4.4.1 Sensitivity of Different Royalty Types to Changes in Commodity Prices: A Simple Comparative Example

As already discussed government needs to balance a number of often incompatible fiscal objectives. While a specific royalty achieves total revenue stability, as long as the mine continues to operate, it has the disadvantage of royalty payments not increasing when/if commodity prices increase.

To a lesser degree this is also the case for value-based royalties, where royalty payments increase modestly as a function of rising commodity prices compared to the significant leverage in company profits.

In both cases public perceptions that companies derive a disproportionate benefit from rising commodity prices creates political pressure for government to change the mining fiscal regime to appropriate a greater share of the rents generated by mining.

These issues could be obviated by adopting a profit-based royalty, although this course of action would sacrifice revenue stability as no royalty would be paid when companies make losses even though their cash flows may be very healthy.

The different sensitivity of royalty payments from different royalty types to changes in commodity prices is illustrated by a simple numerical example.

Let us consider imposing a mineral royalty on a project that produces 1000 tonnes of a commodity currently sold for $100 per tonne with a total cost of production of $80 per tonne. Under current circumstances, a specific royalty of $3 per tonne, an ad valorem royalty of 3% of revenue and a profit-based royalty of 15% would generate, as shown in Fig. 4.2 the same royalty payment of $3000. However, as the price of the commodity rises, as illustrated in Fig. 4.2, royalty payments based on profit are leveraged well in excess of the corresponding revenue generated by the ad valorem royalty, while specific royalty payments remain the same at $3000 and this creates political pressure for a "fairer share" of the profit.

On the other side of the spectrum, when the price of the commodity falls, the specific royalty continues to generate $3000 in revenue, while both the ad valorem and even more so the profit-based royalty collections decrease, with revenue from the latter becoming zero when the commodity price falls below $80 per tonne. Specific and ad valorem royalties will continue to be collected as long as the mine continues to operate. Under this set of circumstances there is no public outcry to give the company a break and government appears generally deaf to any company request for royalty relief unless there is a real chance of the mine closing as a consequence of softer prices. It must be considered, however, that a mine will not close when it becomes unprofitable as, if it has still got significant capital assets to generate depreciation deductions, it may still enjoy very healthy cash flows.

4.4.2 *Hybrid and Multiple-Rates* ad valorem *Royalty*

As discussed above company profits are leveraged more than royalty collections by rising commodity prices and as a result, in single-rate ad valorem royalty systems, total government revenue from royalties decreases as a proportion of the before-tax profit generated by a project. Attempts to overcome the shortcomings of a single-rate *ad valorem* systems, include:

- Hybrid royalty systems which combine a profit-based royalty with a minimum royalty payment in the form of either a value-based or a specific royalty. Hybrids are less economically efficient but more revenue stable than a straight profit system. They, however, present all the administrative complexity which, as already discussed, characterises profit-based systems.
- Multiple-rates ad valorem systems in which the royalty rate increases in steps as a function of increases in the price of the commodity. As exemplified below this approach provides most of the benefits of profit-based royalties but with much lower administrative complexity. The royalty rates for various minerals are published from time to time by the tax authority and are generally calibrated to collect a desired proportion of profits.

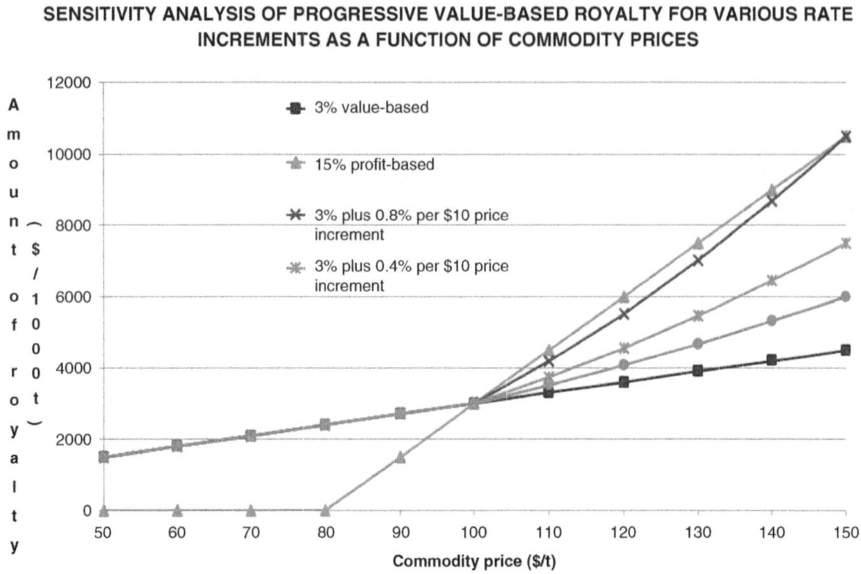

Fig. 4.3 Sensitivity of royalties to variations in commodity price

To maintain its share of before-tax profit against rising prices, the Australian State of Queensland, for instance, applies value-based royalty rates rising incrementally with prices, e.g. for:

- **Coal:** 7% for prices of up to $100/t, 12.5% between $100 and $150/t and 15% above $150/t.
- **Base metals:** 2.5–5% in steps of 0.2% for commodity prices as regularly listed by the Queensland Department of Mines.

On the basis of the previous example for instance, at an expected price of $100/t a 3% value-based royalty and a corporate income tax rate of 30%, the project would pay total tax revenue of $8100 or 47.7% of the project pre-tax profit, including $3000 in as royalty payments.

The diagram of Fig. 4.3 shows how:

- The 47.7% proportion could be roughly maintained by increasing the value-based royalty rate by 0.8% for each $10/t increment in the price of the commodity above some pre-specified expected commodity price level, in the example$100/t.
- The rate of 3% could be adopted as a minimum, so that royalty collections for prices below $100/t are the same as for the single royalty rate approach, that is to say progressively decreasing if the price of the commodity falls but still being received as long as the mine stays open, thus providing a degree of revenue stability.

Figure 4.3 also shows how lower royalty increments result in decreasing proportion of before-tax profit being collected by government. In effect a multirate value-based royalty provides government with a choice as to how much of possible increases in pre-tax profits they wish to capture if/when commodity prices increase above expectations. If desirable the royalty rate could also be capped to a maximum, e.g. 7%.

4.5 Economic Rent Based Royalties

4.5.1 General Background

A resource rent tax (RRT) applies a percentage rate, generally 25–50% to the economic rent generated by a mining or petroleum project.

Economic rent is the surplus after subtracting from the revenue generated by a project all cash costs of production including recurrent and capital costs, but excluding loan interest expenses. In addition a level of normal profit sufficient to attract and retain investment in the project is also allowed as a deduction from revenue in determining the tax base.

There are a number of formulations for RRT including, among others, the Brown (1948) tax and the model originally proposed by Garnaut and Clunies-Ross (1975, 1983), on which most of the RRTs currently in use, primarily within the petroleum industry such as the petroleum resource rent tax (PRRT), are based. This model was also adopted in the past for the mineral resources rent tax (MRRT) which applied for a limited period of time only to iron ore and coal production in Australia, as discussed in more detail below.

Both models are computed on a cash basis, with the taxing point generally set at the well head or mine head. The taxable entity is ‘ring-fenced,’ that is to say tax is levied at the individual project level although some jurisdictions (e.g. Australian PRRT) allow deduction of exploration costs at the holding entity level. Deducting exploration costs at an entity level has, of course, the effect of deferring any RRT payment for many years, particularly if the accumulated exploration costs are uplifted, as in the case of the Australian PRRT, at the long term bond rate (LTBR) plus a high risk premium generously set at 15%.

The Brownian tax model differs in some important respects in that government fully shares project risk with industry by refunding it any losses and as a consequence the applicable normal profit is limited to the risk free rate of interest without the addition of any risk premium.

By contrast in the Garnaut and Clunies-Ross RRT model industry can carry losses forward and uplift them at a rate of interest, that is to say the ‘normal profit’ that, in this case, includes, besides the risk-free rate normally set as the LTBR, a risk premium appropriate to the risk inherent, theoretically in each project, in practice set a unique percentage for the whole industry. The reason for this is that government, while exposed to market and technical risk, does not share in any project losses that may have accumulated by the time the mine closes, and are to be borne entirely by the mining company.

Thus an economic rent based royalty is in theory economically efficient in that, as long as the rate of normal profit is appropriately set relative to the project risk, it does not distort investment decisions or lead to high grading. It is also equitable in terms of the ability-to-pay of different projects in that high-grade/quality mines closer to market will generate higher rents and pay comparatively more tax.

A drawback in terms of practical implementation is that, as already pointed out, normal profit is generally set by policy to be the same for all mining projects. This approach would be valid if all projects had the same level of risk. This is clearly not the case and as a result less risky projects are undertaxed and more risky ones overtaxed. In effect, for a rent-based tax to really be economically efficient different risk premia should be set for individual projects featuring different risk levels or at least for different project categories.

RRT is very complex and costly to administer and comply with in terms of determining allowable expenditure deductions and of its collection that in most cases implies the need to first collect provisional, monthly or quarterly royalties, on an *ad valorem* basis and then, once the extent of the realised economic rent is known after the annual accounts have been audited, a process of reconciliation. Due to its high level of administrative complexity and ambiguity RRT is also open to dispute and prone to avoidance.

In addition, RRT is the most variable in terms of revenue payments, which in many cases may not be received for many years from the start of operations and it makes revenue forecasts for government budgeting very difficult and uncertain. While this revenue instability may not be a worry in highly diversified economies, it can play havoc with planning and budgeting in mineral economies highly dependent of mining taxation revenues.

Furthermore, because RRT is remote from the physical reality of mine production, it is more difficult to explain to the public.

At the time of writing no royalties based on RRT are applied to minerals in any jurisdiction in the world. Resources rent taxes are, however, of common and by and large successful application in the petroleum industry throughout the world particularly for large off-shore projects, but also occasionally to on-shore ones. In Australia PRRT is collected by the federal government both off-shore and on-shore and co-exists with petroleum royalties collected by the states, which, to avoid double taxation, the federal government refunds to industry.

4.5.2 Resource-Rent-Based Royalties: The Australian MRRT Case

A mineral resource rent tax (MRRT) was introduced in Australia in 2012, but limited to iron ore and coal production, at a time of anomalously buoyant commodity prices and of a popular perception that companies' profits were leveraged well ahead of

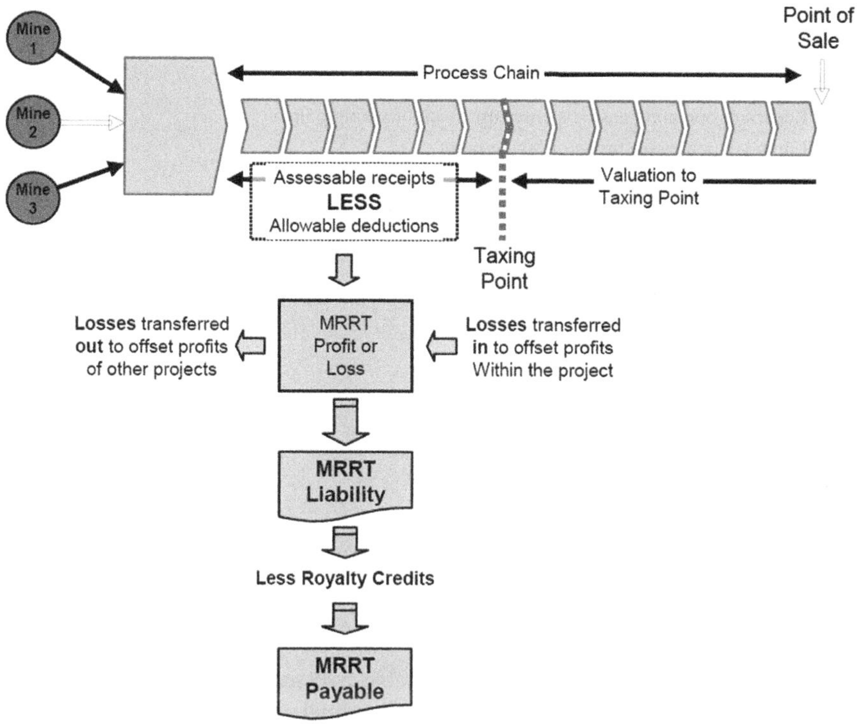

Fig. 4.4 Determination of gross revenue for MRRT in Australia

government revenues and of a need for the community to receive a 'fairer share' of benefit from mining.

The introduction of the MRRT followed a previous proposal for the imposition of a resource super profits tax (RSPT) on all minerals produced in Australia, as originally recommended by the Henry et al. report (Henry et al. 2009), which proved impractical and was strenuously resisted by industry. The final formulation of the Australian MRRT, structured using the Garnaut/Clounie-Ross model, was essentially a poor compromise negotiated by government with only the four major producers on the heave of a federal election under a cloud of secrecy, which planted the seeds of its ultimate demise.

The MRRT introduction also created significant political pressures for other governments to react, including those of many developing countries, by following the Australian example and introduce MRRT in their jurisdictions. However, following falls in commodity prices, poor revenue collection and a change of government in Australia, the MRRT was finally repealed in 2014 and this type of tax is not currently applied to minerals in any jurisdiction in the world.

The taxing point was set at the mine gate. On this account, as schematized in Fig. 4.4 (reproduced from the Australian Government MRRT Exposure Draft and Explanatory Material 2011) determination of the applicable gross revenue at the

Table 4.2 Basic accounting structure for the calculation of MRRT

Gross revenue at taxing point	
Less:	
	Recurrent operating cash expenses upstream of taxing point
	Capital expenses upstream of taxing point
	Unutilised losses brought forward
	Uplift of losses brought forward (**normal profit**) (LTBR + x% risk premium)
	MRRT losses from associated company project
= Assessed taxable economic rent	
Less:	
	Resource rent at a tax rate of y%
= Resource rent tax liability	
Less (in the case of Australia):	
	Royalty credits
= Resource rent tax payable	

taxing point necessitated the netting back of all costs incurred downstream from the mine gate to the point of sale. A project was defined as including all related mine sites in an area particularly if sharing some common mining and transport infrastructure.

For the purpose of determining the applicable assessed taxable economic rent and the MRRT liability, payable cash expenses and normal profit were deducted from the gross revenue at the taxing point, as shown in Table 4.1.

Under the MRRT regime a company was allowed to transfer and defray MRRT losses realised in a project against MRRT profits realised in another project within the same corporate entity.

The picture was highly complicated by the fact that, Australia being a federation, the MRRT collected by the central federal (Commonwealth) government had to co-exist with a range of diverse royalty regimes imposed by the individual Australian States and Territories. Not surprisingly, the states proved disinclined to surrender one of their main taxing mechanisms. Consequently, to avoid double taxation, the federal government had to credit to the mining companies an amount equivalent to their annual royalty payments to the states and territories. This is reflected in the example of Table 4.2 resulting in total tax payments including MRRT and CIT of the order of 44–49% of pre-tax profits, which in the absence of a rebate of state royalty payments would have amounted to a combined mineral royalty plus MRRT and CIT of 51–64%, representing a powerful disincentive to further investment in the industry. A negative spin off of the federal policy of refunding royalties paid to the states to industry, was that it lessened the political pressure from industry that generally inhibits the states from increasing their royalty rates, with some states taking advantage of this peculiar set of circumstances to do so.

The implications of applying the RSPT and its successor MRRT are discussed by Parmenter et al. (2010). Asides from the perceived high rate of tax, this paper identifies the retrospectivity inherent in its introduction as the major issue causing deep disagreement and resistance from industry.

Ergas et al. (2010) discuss the pros and cons of value-based royalties versus profits-based taxes, the issues of RSPT retrospectivity and revenue expectations. Their paper posits that profits-based taxes will discourage resource investment while moderate royalties will not. They further state that the minerals resource rent tax (MRRT) distorted investor returns, thus affecting the attractiveness of a jurisdiction to equity capital investment to develop mineral projects in the first instance.

Hogan (2010) also provides detailed analysis around the proposed RSPT and the MRRT, and compares these tax proposals against profit-based and output-based royalties. The analysis put forward a DCF model that quantified the State accruals, both actual and hypothetical, under the various royalty regimes. He concluded that the MRRT "...will substantially enhance the capacity of Australia's resource taxation framework to obtain a reasonable return from the extraction of the community's resources......"

Guj (2011) demonstrated how the introduction of the MRRT would be biased against smaller and as yet undeveloped iron ore projects in general relative to the handful of major producers that were signatory to an 11th hour agreement with government not to contest its introduction. The main source of bias derived from the generous 'grand parenting' conditions granted by government that allowed current producers to value and depreciate their existing assets on either their book or at market value. As the market valuation option capitalized the value of the vast mineral resources held by existing large producers, they were in a position not to have to pay any MRRT for many years and potentially for decades, while new producers, who were not allowed to capitalize the value of their resources, would have had to start paying it much earlier. In this context, it is worth noting that at the time of the MRRT introduction, Rio Tinto's iron ore assets were conservatively valued at over $110 billion during a frustrated attempt to their being taken over by BHP.

Table 4.3 below provides an example of how the MRRT would be calculated in the case of a typical medium-size iron ore mine located in Australia, in nominal Australian dollars. The calculation is complicated by the fact that both the state royalties and the MRRT are legitimate deductions in computing the taxable income on which corporate income tax is finally levied by the federal government and by the fact that state royalties are credited against assessed MRRT liabilities in determining the final MRRT payable.

From an administrative point of view, the process is further complicated by the fact that the economic rent produced by the mining project cannot be accurately assessed until after the end of the financial year when the company's accounts have been audited. However, to generate a cash flow government has already generally collected provisional MRRT in advance at monthly or quarterly intervals on a value base, which creates the need for a final reconciliation.

Table 4.4 computes the federal MRRT applicable to the iron ore mine example of Table 4.3 and in particular its interplay with the mineral royalties collected by the state in which the mine occurs, which under a federal system such as the Australian

Table 4.3 Assumption used in modelling the MRRT, mineral royalties and CIT of an Australian iron ore mine

Reserves Mt	50
Life of mine years	6
Production rate Mt/y	5 in Y3 and 9 in Y4 to 8
Iron ore price c.i.f. Tianjin $/t	125 down to 115
Initial capital expenses $M (nominal terms)	750
Pre-production period years	2
Sustaining capital $M (nominal terms)	20 Y3 to 5 10 Y6
Mining, crushing and screening cost $/t (real terms)	35
Railing, blending and loading $/t (real terms)	11
Sea freight and insurance $/t (real terms)	13
Nominal cost escalation %	3.5%
Normal profit %	10.5%
RRT Rate %	25%
Corporate income tax rate %	30%
State royalty rate f.o.b. port of export %	7.5%

one are credited back to the mining company to avoid double taxation. The resulting combination of the MRRT and mineral royalty is defined as the 'total resource charge'. The calculation of the 'profit' on which MRRT is on a cash basis and after all capital and operating expenditures and losses carried forward, escalated at the 'normal profit' uplift rate of 10.5%, have been fully deducted from sales revenue. The magnitude of these deductions results in no MRRT potentially being paid for many years after the start of operations.

In the case of the current Australian petroleum resources rent tax (PRRT), which also allows the cost of exploration from different projects to be pooled, carried forward and uplifted before deduction at a real rate of 15% per annum to compensate for the considerable risk inherent in petroleum exploration, deduction of losses brought forward has resulted in considerable deferral of any PRRT payment to the Australian government. The current conditions are considered by government a major weakness of the PRRT system and moves are afoot to amend the related legislative clauses.

Table 4.5, by contrast, displays the calculation of the CIT applicable to the iron ore example keeping in mind that both the MRRT and the state mineral royalties, that is to say the 'total resource charge,' are legitimate deductions in assessing the taxable income of the project. Besides the CIT, this table also displays the 'total resource and fax charge' applicable to the project, both as an absolute amount and as a percentage of profit. It will be noted that the latter is very high (i.e. 57.5%) in the first year of production and then settles within the 44–49% range, which places Australia in the rank of the 'highly-taxing' jurisdictions, negatively affecting, until the MRRT was rescinded, its attractiveness to FDI.

Table 4.4 Financial model for the calculation of the MRRT and mineral royalty of a typical iron ore mine in Australia

Year	**0**	**1**	**2**	**3**	**4**	**5**	**6**	**7**	**8**	**Total**
Resource charge										
Sales Mt				5	9	9	9	9	9	50
Price c.i.f.(Tianjin) $/t				125	125	120	120	115	115	
Sales revenue c.i.f. (Tianjin) $M				625.0	1125.0	1080.0	1080.0	1035.0	1035.0	5980.0
Sales revenue f.o.b. Export port $M as royalty base				552.9	990.7	941.0	936.2	886.1	880.9	5188.0
Sales revenue at ROM pad $M as RRT base				492.0	877.1	823.5	814.5	760.2	750.6	4517.8
Operating expenses at ROM pad $M				194.0	361.5	374.1	387.2	400.8	414.8	2132.4
Capital expenses initial and sustaining (CAPEX) $M		300.0	450.0	20.0	20.0	20.0	10.0			820.0
Royalty fob export port $M				**41.5**	**74.3**	**70.6**	**70.2**	**66.5**	**66.1**	**389.1**
RRT profit/loss before normal profit and losses $M		−300.0	−450.0	236.5	421.4	358.8	347.1	293.0	269.7	
RRT unutilised losses carried forward $M		−300.0	−781.5	−627.1	−271.6	0.0	0.0	0.0	0.0	
Normal profit uplift $M		0.0	−31.5	−82.1	−65.8	−28.5	0.0	0.0	0.0	
MRRT profit/loss $M		0.0	0.0	0.0	149.8	358.8	347.1	293.0	269.7	1418.3
RRT assessed $M		**0.0**	**0.0**	**0.0**	**37.4**	**89.7**	**86.8**	**73.2**	**67.4**	**354.6**
RRT payable after royalty credits $M		**0.0**	**0.0**	**0.0**	**0.0**	**19.1**	**16.5**	**6.8**	**1.4**	
Royalty credits carried forward $M		0.0	0.0	41.5	78.3	59.2	42.7	35.9	34.5	
Total resource charge (royalty and/or RRT) $M		**0.0**	**0.0**	**41.5**	**74.3**	**89.7**	**86.8**	**73.2**	**67.4**	**432.9**

Table 4.5 Financial model displaying the calculation of the CIT and the 'total resource and tax charge' payable by the iron ore mine example under a MRRT fiscal regime

Year	0	1	2	3	4	5	6	7	8	Total
Corporate income tax (CIT)										
Sales revenue c.i.f. (Tianjin) $M		0.0	0.0	625.0	1125.0	1080.0	1080.0	1035.0	1035.0	5980.0
Salvage net of closure and rehabilitation expenses $M										152.3
Operating expenses $M		0.0	0.0	327.1	609.3	630.7	652.7	675.6	699.2	3594.6
Total resource charge (royalty and/or RRT) $M		**0.0**	**0.0**	**41.5**	**74.3**	**89.7**	**86.8**	**73.2**	**67.4**	**432.9**
Depreciation $M				192.5	149.4	117.0	90.3	67.7	50.8	152.3
Profit/loss before tax $M				64.0	292.0	242.6	250.2	218.5	217.6	
Losses carried forwarded $M		0.0	0.0	0.0	0.0	0.0	0.0	0.0	0.0	
Taxable income $M		0.0	0.0	64.0	292.0	242.6	250.2	218.5	217.6	1284.8
Corporate income tax payable $M		**0.0**	**0.0**	**19.2**	**87.6**	**72.8**	**75.1**	**65.5**	**65.3**	**385.5**
Net after-tax profit $M		**0.0**	**0.0**	**44.8**	**204.4**	**169.8**	**175.2**	**152.9**	**152.3**	**899.4**
Operating profit before royalty and CIT		0.0	0.0	105.4	366.3	332.3	337.0	291.7	285.0	
Total resource and company tax $M		**0.0**	**0.0**	**60.7**	**161.9**	**162.5**	**161.8**	**138.8**	**132.7**	**818.3**
Total taxes (resource and CIT) as a percentage of profit %				**57.5%**	**44.2%**	**48.9%**	**48.0%**	**47.6%**	**46.6%**	

4.6 Production Sharing Contract

The name Production Sharing Contract (PSC) is deceiving because, under its most common formulation, as depicted in Fig. 4.5, it is not physical volumes of products, but profits that are in fact shared between a company and government.

PSC sharing is a taxing mechanism commonly used in the petroleum industry but nowadays rarely for minerals.

In the majority of cases the company bears the cost and risk of exploration and development of petroleum projects in exchange for:

- recovery of all costs incurred to bring the project into production generally from the revenue of the project after payment of government royalties; and
- a share of the "profit" from subsequent production on which the company is liable for corporate income tax.

A PSC differs from corporate income tax primarily in so far that the contractor recovers its costs as they emerge.

A PSC is very different from a contract of work (CoW) another arrangement commonly adopted by the petroleum and to a lesser extent by the mining industry. As illustrated in Fig. 4.6, in a CoW the contractor carries out the necessary exploitation work and is remunerated by government for it. Income tax is then payable after deduction of all costs incurred by the contractor generally on an emerging basis.

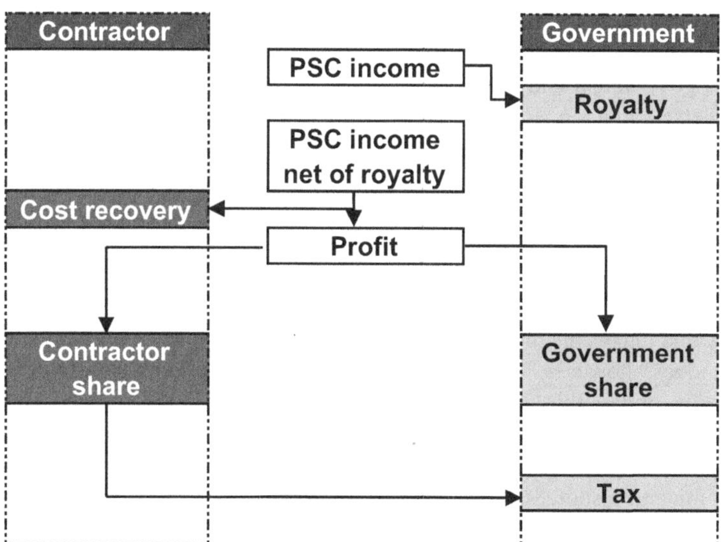

Fig. 4.5 Production sharing contract schematic showing cash flows

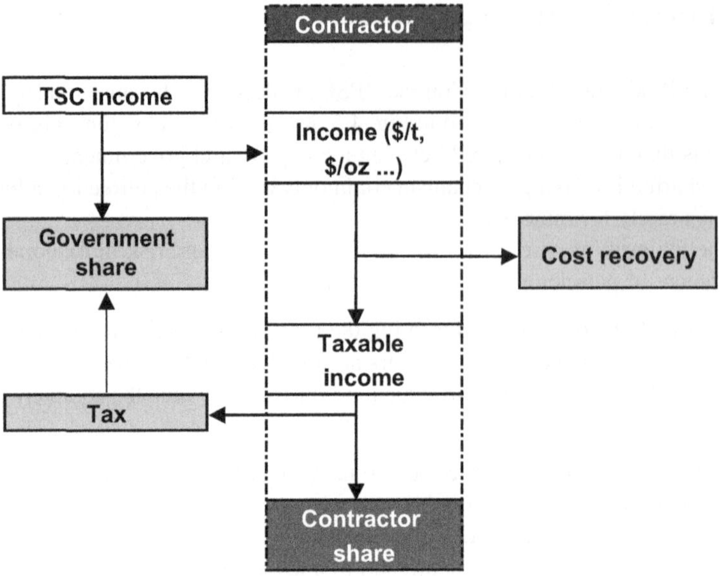

Fig. 4.6 Contract of Work (CoW) schematic showing cashflows

4.7 Concluding Remarks

In summary, all royalty systems have advantages and disadvantages and the government's choice as to which to adopt will always be a compromise to balance a number of at times mutually incompatible objectives. The choice will vary with government's revenue needs and expectations, its competition with other jurisdictions in attracting limited FDI funds, the degree of economic diversification of the country, government and company attitudes and administrative skills and community understanding.

In general, the need for revenue adequacy and stability tends to prevail, particularly in mining, at the expenses of economic efficiency and equity. As a result most jurisdictions make use of production related royalty systems, i.e. specific volume/weight-based and above all value-based royalty systems in preference to profit-based royalties. This choice also achieves the objective of simplifying the related administrative processes and of reducing compliance costs for both government and industry and makes the process more transparent and easy to explain to the community. It does, however, creates political pressure, when following increases in commodity prices, the community develops a perception that royalty collections are not keeping pace with the much more leveraged company profits and demands changes to the system to capture a 'fairer share' of the benefit from the exploitation of their mineral resources.

Governments have also generally shied away from applying economic rent-based and production sharing systems to levy royalties on mineral production and by and large limited the use of these systems to the petroleum industry.

While harmonisation of fiscal systems is generally desirable, some minerals/projects with significant potential economic impacts or of a strategic nature may require unique royalty arrangements and fiscal stability at the cost of increased administrative complexity. If, as discussed later, fiscal stability is achieved through the mechanism of 'stability agreements' then they should incorporate explicit timings for review and 're-negotiation trigger' mechanisms.

Take Away Points

Government's point of view

- A mineral royalty represents the consideration paid by a mining company to the state for the right to mine and sell the latter's non-renewable resources.
- Royalty payments should reflect the value of the resource as taken from the ground, i.e. at the mine head or well head and not on the value subsequently added to mineral products by downstream processing.
- Consequently the royalty rate should gradually decrease if the selling price of a partly or fully processed mineral product (e.g. ore or concentrate or metal) is selected as the royalty value base.
- It is pointless for government to impose requirements for downstream processing by legislation if a business case cannot be built for the individual projects resulting in upstream mining operations cross-subsidising downstream processing.
- While there is little ambiguity about the sales value realised by fully downstream integrated companies selling metals, many royalties are based on the fob value of the mineral product (generally ore or concentrate) at the port of export. In the absence of contractual documentation, this necessitates netting back from the sales value of the contained metals (if readily available) smelting and refining and shipping costs, a process often the subject of disputes.
- The market for other intermediate products can be extremely opaque and sales prices hard to verify in the absence of documentary evidence such as sales invoices. This may create disputes when the sales take place in a foreign jurisdiction and no documentation is made available to the taxing authority. In these instances it would be desirable for the royalty regulations to include an obligation for the company to submit sales invoices or failing it for the Minister to have price deeming powers under the legislation.
- While ad valorem royalties are to some degree economically inefficient, this can be improved by adopting progressive rates as a function of commodity prices. This however introduces a higher level of administrative complexity.

• Profit-and/or economic rent-based royalties are rarely or not currently used for minerals but commonly applied together with product-sharing to petroleum production.

Mining industry's point of view

• Most miners understand and agree with the philosophy that mineral royalties are strictly not a tax and have relatively little difficulty with complying with specific and ad valorem royalty regimes.

• Mining companies should lobby very vigorously not to have a single royalty rate to apply to different mineral products of the same metal and point out that this is a strong disincentive to invest in downstream processing.

• In the long run it pays to be honest and forthcoming with the provision of sales information to the tax authority thus building trust and a cooperative rather than an adversarial relationship with them.

• For projects with significant potential economic impacts or of a strategic nature government may be willing to negotiate unique temporary or permanent royalty reductions and fiscal 'stability agreements'.

References

Australian Government (2011) MRRT exposure draft and explanatory material. MRRT Policy transition Group Secretariat, released June 19, 2011

Boadway R, Keen M (2014) Rent taxes and royalties in designing fiscal regimes for non-renewable resources. CESifo Working Paper No 4568. Category 1, Public Finance

Brown CE (1948) Business-income taxation and investment incentives. Income employment and public policy: essay in honor of Alvin H. Hansen. New York. Norton

Burness HS (1976) On the taxation of nonreplenishable natural resources. J Environ Econ Manag 3 (4):289–311

Conrad RF, Hool B (1981) Resource taxation with heterogeneous quality and endogenous reserves. J Public Econ 16:17–33

Conrad RF (1978) Royalties, cyclical prices and the theory of the mine. Resour Energy 1:139–150

Department of State Development and Department of Mines and Petroleum (of Western Australia) (2015) Mineral royalties rates analysis: Final report 2015. http://www.dmp.wa.gov.au/Documents/Minerals/Mineral_Royalty_Rate_Analysis_Report.pdf

Ergas H, Harrison M\, Pincus J (2010) Some economics of mining taxation. Economic Papers of the Economic Society of Australia, 29(4). pp 369–389

Garnaut R, Clunies-Ross A (1975) Uncertainty, risk aversion and the taxing of natural resource projects. Econ J 85:272–287

Garnaut R, Clunies-Ross A (1983) Taxation of mineral rents. Clarendon Press, Oxford

Gillis SM et al (1978) The Indonesian mining sector: Tax and related policies. HIID, Cambridge

Guj P (2011) Is MRRT competitively neutral? Centre for Exploration Targeting, Newsletter, September 2011

Guj P, Bocoum B, Limerick J, Meaton M, Maybee B (2013) How to improve mining tax administration and collection frameworks: a source book. World Bank/CET, Washington, DC

Hartman F, Guj P (2013) Chapter 12. Mineral taxation and royalties. In: Maxwell P, Guj P (eds) Mineral Economics: Australian and Global Perspectives. 2nd edition. - Monograph N. 29 of the Australasian Institute of Mining and Metallurgy. BPA Digital Burwood, Victoria, Australia

Henry K, Harmer J, Piggott J, Ridout H, Smith G (2009) Australia's future tax system: report to the treasurer. December 2009. Canberra. https://taxreview.treasury.gov.au
Hogan L (2010) Non-renewable taxation: policy reform in Australia. Aust J Agric Resour Econ 56:244–259
Hotelling H (1931) The economics of exhaustible resources. J Polit Econ 39(2):137–175
Laing G (1976) An analysis of the effects of state taxation of the mining industry in the Rocky Mountain states. Masters thesis. Colorado School of Mines
Lilford EV (2017) Quantitative impacts of royalties on mineral projects. Resources Policy 53:369–377
Lockner AO (1965) The economic effect of the severance on the decisions of the mining firm. Nat Resour J 4:468–485
Otto J, Andrews C, Cawood F, Doggett M, Guj P, Stermole F, Stermole J, Tilton J (2006) Mining royalties – a global study of their impact on investors, Government and civil society. The World Bank, Washington, DC
Parmenter B, Breckenridge A, Gray S (2010) Economic analysis of the Government's recent mining tax proposals. Econ Papers: J Appl Econ Policy 29(3):279–291
Smith JL (2013) Issues in extractive resource taxation: a review of research methods and models. Resour Policy 38:320–331
Steele H (1967) Natural resource taxation: resource allocation and distribution implications. Gaffney. pp 233–267

Chapter 5
Corporate Income Tax Provisions and Fiscal Incentives Specific to Mining

Abstract Unlike royalties that, in the majority of cases, may be imposed irrespective of profitability, corporate income tax (CIT) is only charged when an operation generates taxable income, in other words if it is profitable. Most fiscal regimes recognise the unique characteristics and risks of the mining industry and provide a variety of fiscal incentives primarily designed to alleviate them and attract investment to the industry. While some of these incentives are economically efficient, others are hard to justify from the government's point of view on a cost-benefit basis, particularly in cases where they may also elicit undesirable industry behavioural responses.

Keywords Fiscal incentives · Tax relief · Capital recovery · Tax holidays · Rebates

5.1 General Considerations

All mining taxation packages include corporate income tax (CIT) as one of their substantial components. CIT is levied on an entity's taxable income, generally assessed on an accrual accounting basis after deducting from revenue all recurring operating, interest and capital recovery expenses allowable under the prevailing tax legislation.

While CIT is normally assessed on an annual basis, it is generally payable in the form of a series of quarterly, or for large enterprises monthly, 'provisional' tax payments based on on-going or forecast business activities. These are then subject to reconciliation after the annual financial accounts are finalised and audited. Tax returns are invariably submitted by companies on a self-assessed basis and increasingly electronically.

In the majority of jurisdictions, a common tax rate is levied as a proportion of the taxable income irrespective of which sector of the economy generates it. While there may infrequently be minimum thresholds below which no tax is payable, instances of progressive tax rates, while common in the context of personal income tax, are rare in the context of CIT in the commercial arena.

E. Lilford, P. Guj, *Mining Taxation*, Modern Approaches in Solid Earth Sciences 18,
https://doi.org/10.1007/978-3-030-49821-4_5

In a minority of jurisdictions, however, taxable income generated by the mining industry is singled out and taxed at a specific rate, generally higher than that for the rest of the economy. In the case of Ghana, for instance, a differential of 5% in the tax rate, i.e. from 30% to 35%, was selected for the mining industry as an alternative to the introduction of a more administrative complex resource rent tax.

CIT is generally levied at the entity level which, in the mining sector, can be either:

- a national or multinational corporation often operating in disperse locations across multiple fiscal regimes; or
- a project/licence-specific company registered in the country to own and operate a specific mining asset.

The latter practice of assigning resources to one purpose so as to contain their use, known as 'ring-fencing,' is most commonly adopted for mining projects in developing countries. From a government's point of view ring-fencing helps to avoid potential delays in government receiving revenues from a mining project by disallowing reductions in its taxable income from tax write-offs of expenditures incurred beyond the producing area or by an 'associated' company. By contrast, under most jurisdictions expenditures by an associated company would be deductable if the project was co-owned by the same integrated taxable entity or it satisfied the level of cross-ownership as required by the local tax laws.

The benefits of ring-fencing, however, are achieved by government at the cost of creating a disincentive for companies to invest in exploration activities beyond the project area, which can slow the discovery and development of new projects, hence negatively impact the potential growth of the minerals sector of a country.

It is critical when CIT is levied at a project level that clear ring-fencing project definitions are in place, in a manner that is consistent with the over-arching policy, because while ring-fencing of operations is one method of ensuring early government revenue, it can also foster aggressive tax planning.

Taxes are also levied in the event that the disposal of an asset or of a company generates a capital gain by selling that asset for a price higher than its written down value for tax purposes. Not all jurisdictions have provisions for application of capital gains tax (CGT), but the trend is for its progressive introduction in most regimes. Some jurisdictions charge CGT on a concessionary basis at a fraction of the applicable CIT rate. In the event that an asset sale realises a capital loss, however, the vast majority of jurisdictions will not allow the loss to be used as a deduction against a company's recurrent operating taxable income, but generally only against any capital gain.

Frequently in mining, CGT is deferred by means of the rollover of the revenue generated by the disposal of an asset into the acquisition of a new identical or equivalent replacement asset. This is often the case in long lived mining operations where the life of the mine may exceed the useful lives of the majority of the assets that it utilises, which must be replaced to sustain its productive capacity. An asset roll-over may also be allowed if the seller is not ring-fenced and the asset is sold to a related company within an integrated taxable entity.

5.2 Fiscal Incentives as a Strategy to Attract Foreign Direct Investment (FDI)

Although CIT is levied on all sectors of the economy and is not unique to the mining sector, most jurisdictions recognise the unique characteristics of mining including the fact that:

- mineral resources are exhaustible, that is to say not renewable;
- mineral exploration and mining are high-risk activities with no guarantee of discovery or mine development and profitability;
- mine development is capital intensive with long lead times before mining projects generate any revenue;
- mineral prices and exchange rates are highly volatile resulting in mining revenues fluctuating widely over time; and
- mining operations can be long lived.

For these reasons and in an endeavour to attract foreign direct investment (FDI) into mineral exploration and mine development, governments tend to differentiate the treatment of this industry in their fiscal regimes.

A mining fiscal incentive may be defined (Readhead 2018) as any special tax provision granted to the mining industry that provides favourable deviation from the general tax treatment applying to corporate entities in general. Fiscal incentives may be designed to either increase the rate of return on investment for a specific or a category of investments, or to reduce their costs or risks.

As a general rule, from the point of view of industry, fiscal incentives are most effective if they deliver benefits at the early stages of a mine development when projects generate significant demand for capital. A fiscal incentive can provide financial relief and accelerate a company's cash flows in two ways, by either:

- waiving (e.g. tax holidays); or
- deferring tax payments to a later date (e.g. accelerated depreciation and amortisation deductions).

The Intergovernmental Forum on Mining, Minerals, Metals and Sustainable Development (IGF) has recently published a "Mining Tax Incentives Database" (2019) categorising the various types of fiscal incentives and detailing their application in 104 mining contracts across 21 countries. The fiscal incentives are categorised as those applying to mineral royalties, CIT and a host of other relevant imposts.

As discussed later in this chapter some fiscal incentives may also elicit undesirable taxpayer behaviour and there are a number of reasons why, from the point of view of government, waiving of taxes through various forms of tax holidays, while a favourite of industry, may not be a desirable strategy from the point of view of government. By contrast the use of accelerated asset depreciation for the purpose of assessing taxable income presents greater opportunity for generating significant industry benefit at a much lower impact and risk to government revenue.

Fiscal incentives can either be provided for:

- the mining sector in general; or
- an individual mining project.

In the former case, the incentives take the form of sector-specific provisions embodied in the Tax Law. In the latter, they represent contractual clauses, contained in the individual special agreement (for details see Chap. 8 below) under which the project is to be developed, designed to modify or override the CIT, mineral royalties and other imposts normally enforced through the Tax Law, the Mining Act and a number of other laws and regulations.

Fiscal incentives provided to the mining sector in general under the Tax Law rarely result in a reduced tax rate. More frequently they take the form of temporary or permanent exemptions, deferrals or, as discussed later in this Chapter, generous deductibility of specific items of capital expenditure.

By contrast, as discussed in detail in Chap. 8, special agreements may frequently contain temporary or permanent reductions in rates particularly for mineral royalties and less frequently for corporate income tax.

There is a wide range of different opinions as to whether or to what degree various forms of fiscal incentives are effective in attracting FDI. These are comprehensively reviewed in Kraal (2019) who indicates that Holland and Vann (1998), Luo and Yan (2010) and Babajide et al. (2014) contend that fiscal incentives are effective in attracting FDI, while Easson (2001a, b), Easson and Zolt (2002), Zolt (2015), Deasai and Jarvis (2012) and Abimanyu (2016) as well as recent research by the World Bank, the OECD and the IMF contend that they are not.

Even where proving successful, fiscal incentives tend to be subordinate to more fundamental determinants of investment attraction (UN, 2000), such as in the case of mineral exploration and mining the geological prospectivity of the country and the level of available infrastructure and skilled labour within that country. One advantage of fiscal incentives relative to most other fundamental investment attraction criteria is that they are fully under the control of government and they can be implemented and changed quickly. In addition, they do not involve immediate cash outflows for developing countries that are generally starved of public funds.

Nevertheless, while framing a package of fiscal incentives, government must also be cognisant of its efficiency in the form of the expected quantum of FDI attraction benefit per dollar of revenue foregone and of compliance cost both for government and industry. It stands to reason that fiscal and/or other forms of incentives are most beneficial to the country when applied to marginal projects, which would otherwise not progress to development. This requires detailed economic analysis at the level of individual projects to determine which projects truly deserve incentives and would not be progressed on their own merit.

Although the amount of tax revenue to be foregone could create a more effective investment attraction if invested in a different type of incentive, as for instance the provision of infrastructure, this alternative would not necessarily be available to most developing countries because of the need for upfront capital investment. It is in this context that international financial aid should be and is frequently provided to developing countries by international institutions, such as the World Bank and the International Monetary Fund, in the form of soft loans at favourable terms.

Fiscal incentives often encountered in the mining industry include:

- Exploration incentives, captured through:
  - mineral exploration incentive schemes;
  - flow-through shares (see below);
  - tax relief on disposal of exploration tenements;
  - immediate expensing of exploration costs and their deductibility from the taxable income of associated companies (where allowed). See 'special capital recovery provisions' below.
- Special capital recovery provisions:
  - immediate write-off of exploration and/or other capital expenses;
  - accelerated depreciation for mining-specific capital expenses;
  - capital and depletion allowances;
  - R & D concessions and rebates.
- Tax holidays;
- Exemptions from VAT, excises, import-export custom duties/tariffs, capital gain and withholding taxes, fringe benefit tax (FBT); and
- Stability agreements.

More than one of the above fiscal incentives are typically apparent in the various jurisdictions.

5.3 Fiscal Incentives for Mineral Exploration

5.3.1 Mineral Exploration Incentive Schemes

As already discussed, the primary factor in attracting FDI to mineral exploration is creating a perception that a country has a high degree of mineral potential. This perception is generally based on industry continuously re-interpreting the country's growing geological knowledge base and on a history of successful exploration and mine development. It is thus paramount that government should collect and make readily available to industry on a pre-competitive basis all relevant geoscientific information gathered to date to enable industry to formulate and justify its exploration programs and strategies.

A country's geoscientific information base primarily includes information at two different scales, i.e.:

- regional; and
- mineral deposit/mining camp scale.

Collecting the former information is largely beyond the scope of industrial research and is primarily the task of a country's Geological Survey. The latter information, by contrast, will be primarily derived from the exploration and mining work carried out by industry and reported to government, frequently as a condition for the granting of their exploration and/or mining titles and from academic research.

While industry may insist that the information it provides should remain confidential while the related exploration/mining titles are alive, government should insist and regulate on its being collated in an appropriate form and publically released as soon as possible to stimulate further exploration. Effectively managing and integrating the information generated by industry within a comprehensive geoscientific database easily and ideally freely accessible to industry represents one of the major incentives to FDI and therefore justifies public subsidy.

Mineral exploration is carried out in a range of different geological environments including in so-called:

- greenfields: remote and geologically less known regions with difficult logistics and no or minimal established mines;
- brownfields: areas better known from the geological point of view because of previous discoveries and successful mine developments and the establishment of a reasonable level of infrastructure; and
- near-mine: geological terranes in the immediate vicinity of existing or past operations with a high potential for the discovery of smaller but potentially highly profitable satellite orebodies amenable to provide feed for existing processing plants.

In terms of mineral supplies, the largest proportion of metal is extracted from a minority of major to giant discoveries often of high-quality ore. To the extent that major orebodies invariably have extensive 'footprints' they tend to be discovered early in the exploration history of a geological terrane. While this should be adequate incentive for companies to carry out exploration in greenfield or 'frontier' areas, in reality remoteness, lack of infrastructure and the inherent low probability of discovering a giant or a major orebody (i.e. about 1 in 1000 and 1 in 100 respectively) militate against it and most, particularly small and middle-size, explorers tend to invest in proven brownfields opportunities where the probability of discovery is higher (around 1 in 20) even though the size of the target tends to be much smaller. Mitigating the high risk borne by industry and on account of the high economic multipliers that attach to the development of a major mine create a rationale for government to provide incentives specifically for greenfield exploration and by extension for exploration in general.

Opportunities for greenfield exploration arise when a country opens access to its territory for the first time generally creating a rush of interest on the part of explorers wishing to capitalise on the 'first mover' advantages brought about by the relative ease of detection of any outcropping or surficially detectable orebody. Under these circumstances the probability of discovery is high and so is the expected return on the exploration investment, removing the need for government to provide fiscal incentives to attract investment in exploration.

However, after the initial rush, the vast majority of frontier areas remain unexplored because they are concealed under shallow to deep surficial cover which prevents direct detection of mineralisation. Under this set of circumstances, it may be justifiable for government to provide fiscal and other incentives designed to either reduce exploration costs and/or risk that would otherwise be borne by industry inhibiting investment.

For the above reasons a number of governments, through their Geological Surveys provide a range of exploration investment incentives (EIS) to stimulate increased private sector resource exploration, leading to new mineral and energy discoveries funded through general budgetary processes or using funds raised through mining tenement rents. New discoveries in these areas will increase knowledge of the State's geology and resources, and help increase employment opportunities.

EISs typically include:

- encouraging exploration through cover including understanding and characterising the regolith cover, ground and airborne geophysical surveys, e.g. high quality aeromagnetic, radiometric and gravity surveys and deep crustal seismic and magneto-telluric surveys;
- promotion of strategic exploratory and targeted deep stratigraphic drilling through direct support of explorers by co-funding of innovative exploration drilling projects in frontier and other strategic areas;
- 3D imaging prospectivity mapping integrating geochronology and isotopic fingerprinting, with mineral and petroleum systems analysis to develop 3D visualization and geodynamic setting to better predict and locate new deposits. This should be made accessible to industry by means of an online GIS-based mapping tool that allows users to view, query and map various geology, resources and related datasets; and
- promoting strategic research with industry aimed at the rapid transfer of new geoscience concepts, skills and technologies.

There is credible evidence that the exploration incentives schemes, if properly structured and targeted, can be very effective in attracting FDI, facilitating mineral discovery and generating significant economic multipliers. An excellent example is the independent economic impact study into the EIS provided by the Geological Survey of Western Australia, undertaken by ACIL Allen Consulting in early 2015, which in aggregate, as detailed in Appendix C, indicated that for every $1 million invested in the EIS the long-run expected net benefit to the State, in terms of higher Gross State Product, is $23.7 million.

5.3.2 Flow Through Shares

Asides from a handful of major diversified, generally multinational, resources companies, mineral exploration, particularly in its early stages, is carried out by a multitude of small to medium enterprises (SMEs). Most of the SMEs are pure explorers with no source of income and must rely entirely on availability of equity funds to operate. Availability of equity funds, of course, varies widely over time with market sentiment and investors' appetite for high-risk investment opportunities. As a consequence of not having any taxable income against which they can offset their exploration expenses, these expenses are accumulated and carried forward as losses in the hope that one day the company may become a profitable producer and generates enough taxable income to offset them. In reality this is seldom the case because, given the generally low probability of discovery, only few will make a discovery and of these fewer still will be able to raise the finance to transition from discoverer to successfully building and thereafter commissioning a project and entering into commercial production. Furthermore in many instances, exploration SMEs tend to divest equity in their largest discoveries to major companies which are better positioned to raise the significant funds necessary to develop them. The result is that only a minor proportion of exploration expenses generate future tax shields for the exploration companies and/or their shareholders which is a disincentive to explore.

A flow-through share (FTS) is a tax-based financing incentive to induce investors to increase their exposure to risky mineral exploration. A FTS is a type of share issued primarily by a SME exploration company to an investor, on the basis of an agreement whereby the exploration company commits:

- to incur eligible exploration expenses up to an amount equal to that paid by the investor for the shares; and
- to "renounce" to the benefit of the investor/shareholder an amount in respect of the exploration and development expenses incurred, which, for tax purposes, are considered to be the investor's/shareholder's expenses enabling him/her to deduct them as if incurred directly.

FTSs have been allowed by the tax authorities in Canada but many other countries, including Australia, have looked to introduce them at some point in time. The relevant Canadian legislation has been amended over time to cap the cost to government revenue and to reduce initial loopholes and opportunities for abuse. In simplistic term, the mechanism around flow-through shares is such that individual investors can purchase newly, company-issued shares in an exploration company listed on the Toronto Stock Exchange (TSX) or TSX Ventures Exchange (TSXV) and then deduct that specific equity investment amount from their personal taxable income, irrespective of what their personal tax rate may be. This effectively subsidises their investment, attracting additional investment funds in the exploration sector and consequently, potentially increasing the expected equity return on that investment for the investor. It must be acknowledged that an exploration investment presents a risky equity exposure from the outset. The limitation of Canadian flow-through shares is that it is only available to Canadian investors and the exploration activities must be for Canadian projects. Consequently, if for instance a Canadian company listed on the TSX or TSXV issues new equity to its Canadian investors, but has its exploration assets in West Africa, the investors do not qualify for flow-through share relief.

5.3.3 Tax Relief on Disposal of Exploration Tenements

As already pointed out, much of the exploration work is carried out by SMEs, many of which, while efficient and comparatively successful explorers, generally lack the financial and technical capacity to be effective mine developers and producers. Inevitably this state of affairs results in the majority of holders of exploration rights spreading the risks and future costs by entering into a farm-out arrangement in return for cash, a commitment to undertake a specified amount of exploration expenditure, or a combination of both. The actual assignment/disposal of equity in the project may occur:

- at the time of the agreement;
- progressively as commitments are met;
- after all the commitments have been satisfied; or
- at some other time such as at the option of the farminee.

As an alternative, the holder of a prospecting right may agree to assign a portion of any income that might be generated by the property without a transfer of an interest in the property as for instance by means of a royalty based on 'net smelter return.'

Indeed the business model of many exploration SMEs envisages adding value to grass-root or up-stream exploration projects up to the point where they become marketable while still at the advanced exploration or pre-feasibility stage. In this light, it is of paramount importance to the explorer that:

- the Mining Law has provisions to register deals and for trading in exploration and mining tenements; and
- the Tax Law has favourable conditions in terms of the criteria for determining the taxable base for CIT, stamp duty, etc., relating to the margins or consideration realised on disposal of exploration and mining tenements through farm-outs.

Some developing countries seem not to appreciate the role played by SMEs in exploration and favour dealing with large integrated enterprises with documentable capacity to advance a project through the full mining cycle from exploration to development and exploitation. While it is true that many small exploration companies promote their projects for joint venturing or sale without carrying out any meaningful exploration in the field, government's reluctance to deal with smaller but more agile and entrepreneurial exploration companies and to limit the scope of trading in exploration and mining titles can severely limit the range of potentially successful investors. The problem of junior explorers dealing in "real estate" can, however be obviated by imposing, monitoring and auditing "minimum expenditure commitments" as a condition of title.

The amount spent on exploration, as capitalised in the explorer's Balance Sheet, would in the absence of any special provisions, represent the cost of the asset to be deducted from the value of any consideration received at the time of its disposal for the purpose of determining the base on which Capital Gains Tax (CGT) could potentially be levied. Conversely the acquisition cost incurred by the farminee would represent the base for any applicable stamp duty. In both cases these may represent significant figures triggering relatively high taxation liabilities for both parties to a possible deal at a time when the project in question is still subject to high levels of risk.

From a policy point of view government confronts a dilemma. On the one hand a disposal of exploration rights creates an opportunity to levy taxes early in the history of exploration. On the other hand it may create a disincentive in terms of attracting the most appropriate form of equity funding for escalating on-going exploration expenditure, delay progress and encourage inefficient and opaque types of interim arrangements between the parties designed not to trigger a taxable CGT event.

Some jurisdictions, such as Australia, have opted to reduce the taxation impact on the disposal of exploration rights in the context of farm-out arrangements entered into for the purpose of exploration for and discovery of minerals in up-stream 'high risk' projects. In the case of disposal of an interest in a prospecting right under a farm-out arrangement for a commitment to undertake exploration expenditure without any further consideration, the Australian Tax Law applies balancing adjustment provisions, whereby such a transaction is deemed to be a disposal of property otherwise than by sale. This has the effect that the consideration for the disposal is deemed to be the value, if any, of the property at the date of disposal. The time of disposal would generally depend on the terms of the agreement between the parties. In the case of an up-front transfer of an interest in a prospecting right at the wildcat or grass roots stage, the market value at the time of disposal would generally be low if not nil. As a consequence, no CGT event is deemed to have occurred and no CGT and, in some cases, no or very low stamp duty is levied on the transfer of the interest. Discovery or exploration failure subsequent to the date of disposal will not alter, with the benefit of hindsight, what the market-value was at the date of disposal.

It must be emphasised that the farm-out agreement must be for the explicit purpose of carrying out further exploration and not for any form of initial exploitation and that the expenditure commitments should be consistent with this purpose. There have been cases where a farminee carrying out bulk sampling and conducting initial trial mining and processing followed by the sale of the related mineral products was deemed to have progressed to the exploitation stage and denied the CGT exemption.

5.4 Capital Recovery Provisions

5.4.1 Mining Capital Assets and Depreciation Methods

Mining is a capital intensive industry.

Much of the necessary capital assets are acquired and put in place during the **pre-production** period that may last a number of years during which mine development and construction take place before mine production and mineral sales start and any revenue is received. **Sustaining capital** expenditure will continue to a lesser degree during the life of the mine to maintain its productive capacity by replacing assets that have become obsolete, by prolonging their useful life through non-recurrent maintenance or to implement expansion in production or changes in mine design. Finally potentially significant capital expenditure may be required for the final mine site rehabilitation, either progressive or after the mining operation ceases.

As shown in Fig. 5.1, the asset base of a mining project can be classified into two broad categories:

- **Normal depreciable assets**, and
- **Mining specific assets.**

Normal depreciable assets are all those items of plant and equipment that are not unique to the mining industry and which are in common use in other sectors of the economy.

By contrast, as the name implies, mining specific assets are unique to the mining industry and very seldom found in other sectors of the economy. In spite of a tendency for tax authorities to 'harmonise,' that is to say to reclassify many of them as normal assets, in the majority of cases mining specific assets still represent a large proportion of the asset bases of mining projects.

As for all other assets, mining specific assets can be tangible or intangible. Tangible assets include those relating to:

- exploration and prospecting activities;
- mine development and construction, including overburden stripping for open cut mines;
- mining specific plant and equipment;
- offices, workshops and processing plant facilities and buildings, waste and tailing disposal dams (sometimes classified as normal assets);
- water and electricity supply facilities;
- on-lease and near lease accommodation, mess hall, roadworks, airport and other personnel facilities;
- off-lease transportation facilities including rail, harbours and related stockpiling, blending, truck, train and ship loading and unloading facilities; and
- contribution to the establishment or upgrade of common use public infrastructure and community development activities.

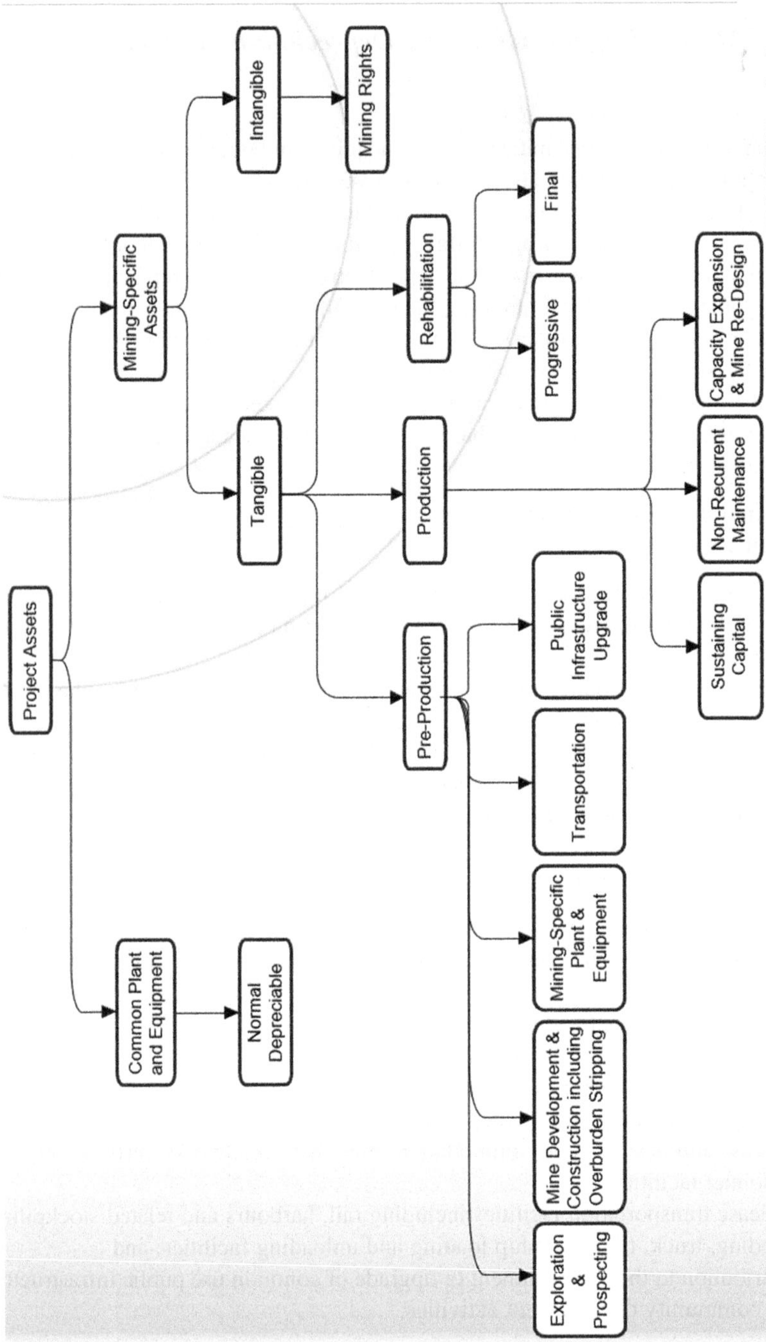

Fig. 5.1 Schematic classification of mining project assets (adapted from Guj et al. 2017)

Table 5.1 Accounting entries relating to the acquisition of an exploration project for a sum in excess of its written down value in the Balance Sheet of the seller

Acquirer limited			
Worksheet for acquisition of target limited			
Details	Assets		
	Cash	Intangible asset—Mining rights	Tangible asset—Exploration expenditure
Cheque—payment to target shareholders for shares in Target	(2,500,000)	1,500,000	1,000,000
Totals	**(2,500,000)**	**1,500,000**	**1,000,000**

A particularly interesting category of intangible assets are the so-called mining rights. Mining rights can be bestowed directly by government most commonly in the form of exploration and mining titles or acquired from another explorer/miner.

Interestingly, while the exploration costs incurred are capitalised, the value of any resources and reserves discovered by an explorer is not recognised as an asset in its Balance Sheet, even though it is the main factor in adding value to the company's shares. The reason is that, due to financial accounting conservatism, the value of the resources and reserves cannot be estimated with an acceptable degree of confidence given their tonnage and grade risk and the volatility of commodity prices and exchange rates. Resources and reserves also fail the ownership criterion as in the vast majority of cases they continue to belong to the state until actually extracted.

However, if following discovery, the relevant mineral resources and reserves are later acquired by a third party for a price well in excess of the capitalised historical cost of discovering them as written down in the books of the seller, the acquiring party's Balance Sheet will show:

- the exploration costs as a tangible asset; and
- the difference between the acquisition price and the exploration cost as an intangible asset (purchased goodwill) referred to as mining rights.

Table 5.1 provides an example of how resources and reserves are capitalised in the Balance Sheet of an acquirer that has paid $2,500,000 for an exploration project on which the seller has to date spent $1,000,000 in exploration.

As discussed in the following sections, the prevailing tax law generally provides detailed procedures and methodologies as to how mining companies should account for and subsequently recover their capital investments in the various categories of assets as a charge against revenue from mineral sales.

Under most fiscal regimes, capital recovery provisions consist in allowing a depreciation and/or amortisation charge against revenue in recognition of the progressive reduction in value of assets as a consequence of their use in generating revenue. The tax law generally specifies the **fiscal useful life** for different categories of assets for taxation purposes, which can range from a few years to many decades. The resultant depreciation charges are then subtracted, together with all other allowable expenses, from revenue to determine the taxable income and annual income tax liability for the project. Some mining specific capital items are depreciated or amortised using their fiscal life or the life of the mine whichever is lower, while others are recovered on a **unit of production basis**. An example could be the depreciation of the expenses incurred in pre-production stripping in an open cut mine or the amortisation of the related mining rights both of which relate to and must be distributed in the context of the extraction of the whole of the resources.

The fiscal useful life of assets may be different and, in most cases, shorter than the corresponding **engineering useful life** which is the length of time during which the assets are actually utilised in the mining operations, as determined on the basis of when it becomes financially advantageous to replace rather than continuing to maintain an asset.

The resultant, generally lower, annual depreciation charges are used, together with the above tax liability and all other accrued revenues and expenses, to determine the annual profit of the mining project as recorded in the Profit and Loss Statement, better reflecting the financial performance of the project for use by external stakeholders.

The fiscal rules may also specify or provide a choice of methods to be used to depreciate various categories of tangible assets and amortise intangible ones. These methods fall within two broad categories, i.e. the:

- **Prime or straight line method**; and
- **Diminishing value or declining balance** method.

Under the straight line method, the amount of annual depreciation is computed by dividing the acquisition cost of the asset by its useful life which generates constant amounts of depreciation over the mine life. The calculation may be conducted in a manner whereby at the end of its life an asset is fully depreciated, i.e. its residual value becomes zero, or it may be computed assuming that the asset will still have a residual value at the end of its useful life.

Under the diminishing value method, the amount of depreciation is computed by dividing the opening written down value of an asset by its useful life. The amount of annual depreciation is then subtracted from the opening written down value to obtain the closing written down value for the year. This will in turn be used as the depreciable value base for the following year and so on. The result is that depreciation is highest in the first year and progressively declines thereafter over the life of the asset which, at the end of its life, maintains some residual value.

Thus relative to the strait line method, the declining balance method 'accelerates' the rate of depreciation and in doing so increases, relative to the straight line method, the project cash flows in the early years of production which can make a significant contribution to the financial success of a project.

5.4.2 Special Capital Recovery Provisions

As already pointed out, special capital recovery provisions, essentially various forms of accelerated depreciation, differ from tax holidays in that they defer but do not waive a mining company's tax liability. The beneficial effects of tax deferrals can be very significant for the success of a mine development, while not as onerous and risky as any form of tax holiday from the point of view of government. For this reason, as discussed in detail below, the authors favour the use of accelerated depreciation, but not of a resource depletion allowance, not only as a potentially effective incentive to attract FDI, but also as a key ingredient in supporting successful mineral development during their critical early years of operation.

Special capital recovery provisions may include one or more of:

- a **capital allowance** based on accelerated depreciation;
- immediate **write-off** of some or all capital expenses;
- **depletion** allowances; and
- **R & D (research and development) concessions and/or incentives**.

A government policy relating to accelerated depreciation targeted to mining specific assets can be implemented at two levels, i.e. by:

- specifying in the tax legislation shorter than normal fiscal useful lives for individual categories of mining assets; and
- applying from time to time a **capital allowance premium** designed to increase the depreciable asset base to create an incentive for industry to invest in new capital assets, particularly at times when the economy is in need of stimulation.

Table 5.2 compares the depreciation charges generated by the two depreciation methods for a mining asset with an acquisition cost of $20 million and a fiscal life of

Table 5.2 Comparison of the depreciation charges generated by the straight line and diminishing value depreciation methods for a mining asset with an acquisition cost of $20 million and a fiscal life of 5 years

	Straight line method			Diminishing value method		
Year	Written down value	Depreciation	Tax shield	Written down value	Depreciation	Tax shield
0	20.00			20.00		
1	16.00	4.00	1.20	12.00	8.00	2.40
2	12.00	4.00	1.20	7.20	4.80	1.44
3	8.00	4.00	1.20	4.32	2.88	0.86
4	4.00	4.00	1.20	2.59	1.73	0.52
5	0.00	4.00	1.20	1.56	1.04	0.31
6	0.00	0.00	0.00	0.00	1.56	0.47
Total		20	6		20	6
Mine life	5	Tax rate	30%	Capital allowance		200%

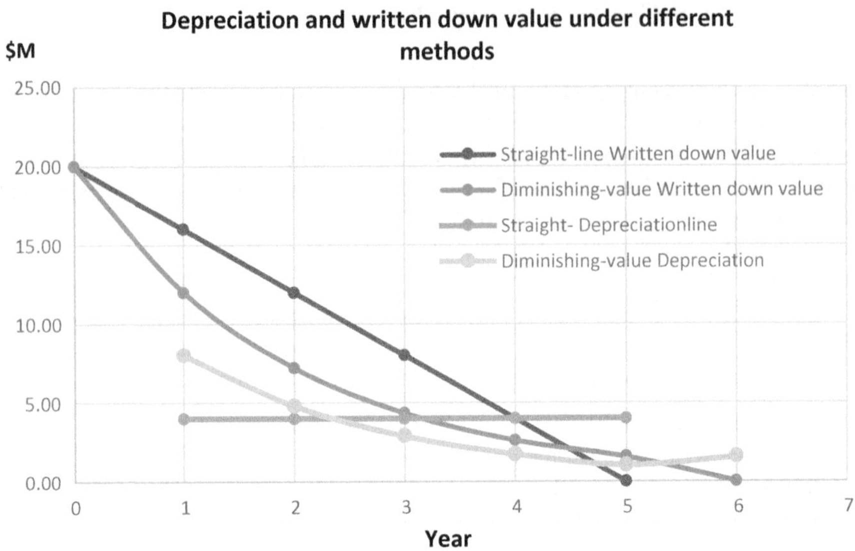

Fig. 5.2 Graphical display emphasising the differences between the annual depreciation charges generated by the straight line and diminishing value methods and their effect on the corresponding written down value of the asset

5 years. In the example, the degree of acceleration under the diminishing value method has been further increased by the government providing a 200% capital allowance. The different pattern of depreciation charges is further emphasised graphically in Fig. 5.2.

Table 5.2 also provides a comparison of the tax shields provided by the two methods during the mining operations which shows how the accelerated depreciation in year 1 at $2.4 million is twice as large as the corresponding straight line one at $1.2 million. It will also be noted that in the former at the end of the mine life the asset still has a residual value of $1.56 million. For the sake of simplicity it has been assumed that the asset is salvaged in year 6, i.e. in the year following the closure of the mining operations, for a price equivalent to its written down value. That is to say that the whole value can be written off without generating a capital gain. In reality the salvage value may be higher or lower than the written down value of assets in which case either a capital gain or loss may be generated and taxed or credited accordingly.

Thus while, as shown in Table 5.2, the amount of accumulated depreciation at the end of the project life will be the same for the two methods, i.e. $6 million, use of accelerated depreciation by creating greater tax shields in the early years of operations will have the effect of deferring taxes to later years thus increasing the project cash flows in the critical early years of operations at the time when a project has the strongest need for capital. To the extent that future tax shields from depreciation are reduced, future taxable incomes are higher leading to tax payments higher than they would have been under a normal, non-accelerated depreciation regime. In effect, accelerated depreciation defers but does not reduce the nominal and undiscounted amount of tax paid.

The net present benefit in the hands of the mining company is enhanced in so far that the unpaid tax will accrue compound interest at the company's discount rate of say 12.5% per annum. This would amount to a total by end of year 6 of $3.74 million for the straight line approach and $4.50 million for the declining value method.

There is also the benefit derived by the fact that depreciation is calculated on the basis of the historical cost of the assets which are carried forward in nominal dollars and its early deduction prevents inflation progressively eroding the degree to which it would shield taxes.

The value of these deferrals and of increasing the project cash flows in the early years of operations is greater in the hands of the mining companies than the opportunity cost incurred by government in providing this incentive because of the significant difference between the cost of capital of the mining company (i.e. 12.5%) as compared to the social rate of discount of government that on the basis of current returns on government bonds is of the order of 2.5%. By way of comparison the cumulative opportunity cost to government of the depreciation deductions by year 6 would be $0.63 million for the straight line method and $0.73 million for the diminishing value one, relative to the already quoted industry's benefits of $3.74 and $4.50 million respectively. On the basis of this significant cost differential between industry and government, it could be validly argued that, on the likely assumption that accelerated depreciation represents a strong incentive in terms of attracting FDI, this fiscal policy presents an element of win-win for both government and industry.

Depletion is primarily a tax and an accounting concept. As the term implies, it may be applied in instances where the produced item comes from a finite source of that item, such as mining, quarrying and oil and gas production. The principles associated with depletion are similar to those that apply to depreciation in that it represents a cost recovery system for accounting and tax determination. The actual deduction allowed by depletion provides the operation's owner or operator to account for the finite nature of its activity, being a continual depletion of the remaining reserves.

A **depletion allowance** is a fiscal incentive to invest that is justified on the basis that the orebody reserves are non-renewable and consequently reduce in size over time as a result of being extracted and sold. The general rules associated with natural resources extraction accounting require that the overall cost expended to find, develop and then exploit the mineral-producing property must firstly be capitalized. Since depletion is a means of depreciating the mineral resources, it allows for a reduction in these capitalized costs from taxable income to adequately cater for the diminishing production reserves remaining as time progresses.

There are three common forms of depletion allowance:

- related to the acquisition price for the resources;
- calculated as a percentage of gross sales revenue; and
- related to the tonnage and value of the ore in the resource supporting mining.

For mining and other natural resources projects, the owner generally uses the method that provides the highest deduction.

In the case of percentage depletion, the actual determination is made by multiplying a percentage, which is typically specific for each mineral type, by the gross income from operating activities over the duration of the tax year. Percentage depletion is often provided as an incentive for explorers and mining companies to develop domestic mineral and oil and gas projects. It is effectively an allowance granted to mineral producers and royalty owners who can then apply the offsets to their project's taxable income.

For cost depletion, the costs associated with the exploitation of the mineral are allocated to a depletion schedule proportioned to the life of mine or operation. The cost depletion amount is calculated by determining the total volume of mineral sold and then attributing a proportionate amount of the total resource cost to the quantity extracted and sold over that period. By way of an example, if a mining company discovers a large reserve of copper, it will draw up a life of mine plan that depletes this reserve over a finite period. If the company plans total extraction of, say, the full 500,000 tonnes of copper and it invests $1,000,000 to extract the copper at 40,000 tonnes in the first year, the depletion deduction is $80,000 ($1,000,000 × 40,000/500,000).

In general, there is no monetary limit on the total amount of depletion that can be deducted for tax purposes. However, percentage depletion can only be determined and used for a project that generates profits. If a project generates a loss over a specific tax year, then the percentage depletion amount calculated cannot be deducted. Also generalising, a percentage depletion is typically limited to an amount equating to 50% of net income, less exploration costs.

It is important to note that cost depletion amount for accounting purposes may be totally different to the cost depletion for tax purposes.

Naturally, care must be taken to avoid a double deduction, which is then, in essence, tantamount to being provided a subsidy.

A depletion allowance tends to be tough to administer as it requires a deep knowledge of the mining industry, which may not be readily available within a government's tax office.

R & D concessions allowed companies to claim a tax deduction in their income tax return in excess of 100%, often up to 200%, of the expenses incurred in carrying out eligible expenditure on R & D activities. Alternatively refundable tax offsets in excess of the prevailing tax rate (say 50% instead of 30%) may be provided as a tax incentive to encourage more companies to engage in R & D.

To be eligible an activity must meet the legislated definition of a core R & D activity which implies that at least one hypothesis-guided experiment is undertaken to generate new knowledge. However other activities supporting the experiment may also be eligible.

Mining strives on innovation with increasing level of research involving development of new technology to explore for blind orebodies under thick cover, remote automation of mining etc. and not surprising is a large beneficiary of these R & D tax incentives where they are provided.

There has, however, been a tendency for some in the mining industry to occasionally claim as R & D expenditure incurred in carrying out tests that are standard to the industry and not hypothesis-driven and consequently to become the focus of specific inquiries and audits from the tax authorities. To avoid adverse tax findings, it is cautious to identify and document budgetary approval of R & D activities and associated expenditure adequately and contemporaneously.

5.5 Tax Holidays

5.5.1 General

A tax holiday is a form of fiscal incentive whereby a company holding a specific mining project will not pay any corporate income tax (CIT) and sometimes mineral royalties and generally fully captures the whole of the profits in the early years of operations.

Tax holidays can be constructed in two general forms:

- on a time basis (e.g. covering the first 5 years of production); or
- on a quantity extracted basis (e.g. covering specific tonnages of ore to be extracted often as is scheduled in the feasibility study);

The analysis that follows makes use of a simplified financial model (Table 5.3) of a gold mine to show the impact that different fiscal incentives can have on the amount of CIT paid. The 'base case' mine has diluted mineable reserves of one million tonnes (Mt) of ore at an average grade of 3.2 grammes per tonne of gold (g/t Au). The real price of gold is expected to average \$1300 per ounce of gold and mining costs are estimated at \$80 per tonne of ore over the life of the mine. Financial figures are expected to escalate in nominal terms at a rate of 3 per cent per annum.

Table 5.3 Simplified "base case" financial model of a gold mine in the absence of any fiscal incentive and applying the straight line method of depreciation

Year	0	1	2	3	4	5
(\$ Million in nominal terms)						
CAPEX	−20					
Production (Mt)		0.2	0.2	0.2	0.2	0.2
Average recovered grade (g/t Au)		3.2	3.2	3.2	3.2	3.2
Revenue		27.55	28.38	29.23	30.11	31.01
Operating expenditure		−16.48	−16.97	−17.48	−18.01	−18.55
Depreciation straight line method		−4	−4	−4	−4	−4
Operating income before tax		7.07	7.40	7.75	8.10	8.46
Corporate income tax		**−2.12**	**−2.22**	**−2.32**	**−2.43**	**−2.54**
Profit after tax		4.95	5.18	5.42	5.67	5.92
Total base-case tax using straight-line depreciation				−11.63		

Table 5.4 Accelerated depreciation method

Year	0	1	2	3	4	5	6
($ Million in nominal terms)							
CAPEX	−20						
Production (Mt)		0.2	0.2	0.2	0.2	0.2	
Average recovered grade (g/t Au)		3.2	3.2	3.2	3.2	3.2	
Sales revenue		27.55	28.38	29.23	30.11	31.01	
Salvage							1.56
Operating expenditure		−16.48	−16.97	−17.48	−18.01	−18.55	
Depreciation diminishing value method		−8.00	−4.80	−2.88	−1.73	−1.04	−1.56
Operating income before tax		3.07	6.60	8.87	10.37	11.42	0.00
Corporate income tax		**−0.92**	**−1.98**	**−2.66**	**−3.11**	**−3.43**	**0.00**
Profit after tax		2.15	4.62	6.21	7.26	8.00	0.00
Total base-case tax using accelerated depreciation and 200% capital allowance						**−12.10**	

The rate of corporate income tax is assumed to be 30 per cent and, for the sake of simplicity, it is assumed that all assets are depreciated fully on a straight line basis over the life of the mine and that no mineral royalty applies.

In the absence of a tax holiday, as shown in Table 5.3, this project would pay a total of $11.63 million in CIT.

Table 5.4 portrays the same model but using the accelerated depreciation method including a 200% capital allowance.

Again in the absence of a tax holiday, as shown in Table 5.4, the same project would pay a total of $12.10 million in CIT.

5.5.2 *Time-Based Tax Holidays*

Let us now assume, as illustrated in Table 5.5, that the mining company has been granted a two-year tax holiday. This means that if the mining company adheres strictly to the production scheduled in the feasibility study, the quantity of ore or metal mined during the first two years will be the same as for the 'base case' but the tax rate for years 1 and 2 is set at 0% whereby no corporate income tax is levied in those years. Under these circumstances the corporate income tax paid will be reduced by $4.34 million from a total of $11.63 million in the base case to $7.29 million.

However, if the tax holiday was to be set on time alone, i.e. if no income tax was to be levied for two years irrespective of the actual rate of ore produced, companies could be tempted to increase the rate of extraction, bringing within the first two tax-free years production that would otherwise have occurred in later years and which would have generated taxable income in the future. An outcome of such a

Table 5.5 Fiscal incentive effect of a two-year tax holiday without any change in production schedules

Year	0	1	2	3	4	5
($ Million in nominal terms)						
CAPEX	−20					
Production (Mt)		0.2	0.2	0.2	0.2	0.2
Average recovered grade (g/t Au)		3.2	3.2	3.2	3.2	3.2
Revenue		27.55	28.38	29.23	30.11	31.01
Operating expenditure		−16.48	−16.97	−17.48	−18.01	−18.55
Depreciation s.l.		−4	−4	−4	−4	−4
Operating income before tax		7.07	7.40	7.75	8.10	8.46
Corporate income tax		**0.00**	**0.00**	**−2.32**	**−2.43**	**−2.54**
Profit after tax		7.07	7.40	5.42	5.67	5.92
Total base-case tax	−11.63	Two-year Holiday		−7.29	**Incentive**	**4.34**

Table 5.6 Fiscal incentive effect of a 25% increase in the rate of production in the first two tax-free years

Year	0	1	2	3	4	5
($ Million in nominal terms)						
CAPEX	−20					
Production (Mt)		0.25	0.25	0.25	0.25	0
Average recovered grade (g/t Au)		3.2	3.2	3.2	3.2	0.0
Revenue		34.44	35.47	36.54	37.63	0.00
Operating expenditure		−20.60	−21.22	−21.85	−22.51	0.00
Depreciation s.l.		−5	−5	−5	−5	0
Operating income before tax		8.84	9.26	9.68	10.12	0.00
Corporate income tax		**0.00**	**0.00**	**−2.90**	**−3.04**	**0.00**
Profit after tax		8.84	9.26	6.78	7.09	0.00
Base-case tax	−11.63	Increased production		−5.94	**Tax Incentive**	**5.69**

strategy is that, while the same amount of ore as in the 'base case' will be mined over the life of the mine, the life of the mine itself will be shortened.

Table 5.6 portrays a situation where the production rate is increased by 25 per cent, from 0.20 to 0.25 million tonnes of ore per annum with consequent reduction in mine life from 5 to 4 years. Under this set of circumstances the total income tax paid by the project would be $5.94 million, a reduction of $5.69 million relative to the $11.63 million total tax that would be paid under the 'base case' or $1.35 million less than the $7.29 million expected under the plain two-year tax holiday scenario.

Alternatively, in the case of a time-based tax holiday, a mining company may seek to minimise its tax payable by increasing the amount of unit profit generated during the first two-year tax holiday period. This may be accomplished by increasing the unit value of the ore mined in the first two years by selectively mining high-grade ore blocks first. This practice, often referred to as 'high-grading,' depends on the mine design having the flexibility to change the production scheduling to access ore

Table 5.7 Fiscal incentive effect of 'high-grading" production during the first two tax-free years

Year	0	1	2	3	4	5
($ Million in nominal terms)						
CAPEX	−20					
Production (Mt)		0.2	0.2	0.2	0.2	0.2
Average recovered grade (g/t Au)		3.55	3.55	2.90	2.90	2.90
Revenue		30.57	31.48	26.49	27.28	28.10
Operating expenditure		−16.48	−16.97	−17.48	−18.01	−18.55
Depreciation s.l.		−4	−4	−4	−4	−4
Operating income before tax		10.09	10.51	5.01	5.28	5.55
Corporate income tax		**0.00**	**0.00**	**−1.50**	**−1.58**	**−1.67**
Profit after tax		10.09	10.51	3.50	3.69	3.89
Base-case tax	−11.63	High-grading case tax		−4.75	**Tax Incentive**	**6.88**

blocks in a chronological order other than the one originally determined as being financially optimal in the 'base case' on which the feasibility of the project was based. A consequence of high-grading is that the ore to be extracted after the tax holiday has a grade lower than average. This negatively affects the project taxable income when the project starts to be taxed from year 3 onwards.

Table 5.7 portrays an example of high-grading production in the first two years when average grades are increased from 3.2 g/t Au to 3.55 g/t Au at the expense of the following three years during which the grade falls from an average of 3.2 g/t Au to 2.9 g/t Au. Under these circumstances, while the total amount of gold produced over the life of the mine remains the same as for the base case, the bulk of it is produced over the first two years. To the extent that profitability in the year 3 till 5 period has been reduced because of the lower unit value of production, the project would pay a total of $4.75 million in income tax That is to say a reduction of $6.88 million relative to the $11.63 million of the 'base case' or $2.54 million less than the $7.29 million expected under the plain two-year tax holiday scenario. By reducing grades in later years, high-grading also increases the vulnerability of the project to potential falls in the price of gold which, in effect increases its risk of closure in later years.

5.5.3 *Tax Holidays Based on Quantity of Ore or Metal Extracted*

To avoid the tax consequences of unspecified time-based tax holidays, a tax holiday may be based on the tonnes of ore to be mined or, alternatively, the metal recovered during the tax-free period, as forecast in the mine model supporting the feasibility study. The latter approach naturally discourages high-grading.

Even in the case of tax holidays based on the quantity of ore or metal extracted, some tax avoidance may arise depending on whether the fiscal regime clearly

Table 5.8 Fiscal incentive effect of allowing accumulation and carry forward of depreciation charges incurred in the first two tax-free years

Year	0	1	2	3	4	5
($ Million in nominal terms)						
CAPEX	−20					
Production (Mt)		0.2	0.2	0.2	0.2	0.2
Average recovered grade (g/t Au)		3.2	3.2	3.2	3.2	3.2
Revenue		27.55	28.38	29.23	30.11	31.01
Operating expenditure		−16.48	−16.97	−17.48	−18.01	−18.55
Depreciation s.l.		0	0	−12	−4	−4
Operating income before tax		11.07	11.40	−0.25	8.10	8.46
Corporate income tax		**0.00**	**0.00**	**0.00**	**−2.35**	**−2.54**
Profit after tax		11.07	11.40	−0.25	5.75	5.92
Total base-case Tax	−11.63	Depreciation Carry		−4.89	**Incentive**	**6.74**

specifies that the tax rate during the tax holiday period is deemed to be zero per cent as opposed to the operation cumulating and carrying forward depreciation during the tax-free period for subsequent deduction from taxable income produced in later years. This may have a significant effect on the income tax collected in jurisdictions allowing the carrying forward of previous years' losses, particularly if they also allow accelerated depreciation.

Table 5.8 portrays how, under these circumstances, the project would pay a total of $4.89 million in corporate income tax. That is to say a reduction of $6.74 million relative to the $11.63 million of the 'base case' or $2.40 million less than the $7.29 million expected under the plain two-year tax holiday scenario. This would be equivalent to extending the tax holiday from the originally intended two years to three years instead.

5.5.4 How to Avoid Tax Holiday Pitfalls

To ensure the tax holiday is not being abused, government should cast the legislation in a manner that clearly defines:

- what constitutes a mining project. Complexities may arise if/when a satellite orebody is discovered and developed, both within and adjacent to the original mining leases. Government will tend to view such developments as an extension of current operations. By contrast, mining companies will often argue that the new operation is separate from that covered by the original mining agreement and should trigger a new tax holiday;
- the extent to which quantities of ore extracted and/or metal recovered can be changed during the tax holiday period;

- how depreciation expenses and losses incurred during the tax holiday period are to be treated in terms of carrying them forward as legitimate deductions against future project taxable income; and
- clarity on for how many years a loss can be carried forward .

Past experience indicates that, more often than not, tax holiday conditions have been drafted in developing countries without a full understanding of their potential future consequences. This created the basis for administrative misunderstanding, legal arguments and litigation. To avoid this, it is critical that future provisions for tax holidays be carefully and specifically drafted to avoid undesirable loopholes and are clearly explained to and fully understood by both parties. In essence, clearly understanding the fiscal regime of the country in which the project is located is imperative for potential mining investors. This is not only to ensure that the operations remains compliant, but it is of equal importance earlier in the process when negotiations are underway between the operators and the government, for the company to secure the best possible tax outcome and minimise potential future controversies and litigations.

On balance, however, unless a tax holiday is an inevitable condition for a mine development to go ahead, it may be prudent for government to consider other tax and non-tax incentives, recognising that tax holidays may cause administrative complexity, ambiguity and potential for tax avoidance and litigation.

5.6 Other Fiscal Incentives and Exemptions

Certain taxes and royalties are not negotiable, but certain concessions and possibly even exemptions may be negotiated on other taxes and imposts. Notably import and export taxes may be waived if the government is convinced that the future economic and other benefits associated with production operations significantly outweighs the menial income that may be derived by the government through import duties and even export duties or taxes.

It is also possible to negotiate a moratorium on royalties for, say, the first five years of production. However, if this is acceptable, it is incumbent on the operator to not high-grade the deposit during the moratorium period, as tempting as that may be.

5.6.1 Exemption from Value-Added Tax (VAT)

A value added tax (VAT) is based on a taxing method that was introduced into the world of taxes in the early 1950s. The primary purpose supporting its introduction is that it is a tax levied against every buyer of a product or ore in the minerals sector, all the way up the supply chain, from the first purchase of raw materials through to the end retail consumer of the finished product. VAT also applies to the acquisition of services.

Many countries around the world, including within the European Union (EU), have a VAT or general sales tax (GST). VAT has been described as:

- a general tax that may apply to every commercial activity that involves the production and sales of products and the provision of services;
- a consumption tax since it is ultimately carried by the product's final consumer. It is not intended as a charge on businesses.

Some goods and services are exempt from VAT and/or GST. If the goods or services sold by the company are exempt, that company is exempt and will not be able to register for VAT. This means that that company cannot reclaim any VAT on its purchases or expenses.

Mining operations purchase numerous items, supplies and services, so are typically registered for VAT. On the sales side, these same operations may sell either crude ore, a partially processed intermediate product, or a completely refined final product, which may be a metal. While the producer and seller of these minerals or materials does not pay VAT, the purchaser of these materials, being a concentrator, refinery or fabricator, will. This effectively increases the cost of that product and is why many mining companies negotiate an exemption from VAT before they make their initial capital investment to develop an operation. An exemption, for both purchases and sales, may be negotiated for a finite term or for the life of the operation.

Furthermore, if the company is VAT-registered and does incur VAT on items that will be used to make VAT-exempt products, that company will be classified as partly exempt.

Exempt items are not the same as zero-rated products. In each case, VAT is not added to the sales price, with the specific difference being that zero-rated goods or services are taxable for VAT, but at 0%.

5.6.2 Exemption from Import-Export Custom Duties/Tariffs

A duty, which is often referred to as a tariff, is a tax levied by governments on the value of many imported or exported products, and also includes activities such as freight and insurance. Varying countries apply different duty rates against different products.

National sales taxes as well as local taxes and potentially customs fees, may be charged in addition to the duties, the latter being collected at the point of customs clearance in the domestic port, whether for export purposes or for import purposes.

As can be expected, duties and taxes increase the cost of the sold product to the offshore buyer and the imposition of duties will typically reduce the company's competitiveness in the market. Therefore knowing what the final cost will be due to additional sales or export duties to the purchaser of the products will help the seller price its product for that market. This assumes that the seller has control over the price of the product, which may not be the case in the minerals industry.

Import duties, also known as customs duties, tariffs, import taxes or import tariffs, have two distinct purposes:

- they generate additional income for the government; and
- they provide a fiscal incentive to locally produced goods that are not charged import duties.

A third purpose may also be to penalize a particular country by charging high import duties on its exported products and so discourage either dumping or excessive "economies of scale" sales.

Specifically in the mining industry and especially in the early years of a project's construction, significant material will be brought into the country for the construction. Most companies will negotiate an import exemption for this new incoming capital equipment, which is an understandable and fair position since the value of the construction material and operating equipment being brought into the country tends to be very high. The companies may also negotiate an export duty exemption for their finished goods, motivated by the fact that they will probably be paying taxes and royalties more than sufficient to render these duties somewhat superfluous and administratively challenging.

A free trade agreement (FTA) may exist between countries whereby import and export duties are either reduced or are totally eliminated.

An excise tax, on the other hand, is a tax that is taken in the same way that a royalty is deducted from operations. Typically, excise taxes are based on a unit of production at a specific rate, such as $0.50 for each tonne of coal produced. They are not dependent on whether the product is exported or imported, but merely on whether they are produced. They are also sometimes allocated to communities or other groups and not necessarily to the government, although they are often collected by the government on their behalf. Exemptions from excise taxes may also be negotiated.

5.6.3 Rebates on Fuel and Other Excises

Many jurisdictions charge excises on the domestic production of certain items considered to impose a community cost. For example excise may be levied on petrol and diesel fuel on the basis that it is a means to recover from road users the costs they impose on society, such as wear and tear on road pavements. Fuel excises may also in theory be justified to reduce negative externalities created by road users, that is, the unpriced costs of road transport such as traffic congestion, road accidents, and noise and air pollution, although setting rates that reflect the cost of these externalities would be difficult.

Some jurisdictions rebate the fuel excise implicit in the fuel consumed in carrying out certain activities including mining on the basis that its use of fuel is primarily off public roads.

Excises are generally legislated as a quantum amount per volume and generally indexed to maintain their value relative to the changing price of fuel over time.

5.6.4 Exemption from Withholding Taxes

The rapid trend in increasing globalisation of the mining industry has resulted in many mining projects in developing countries being owned and operated by subsidiaries of multi-national enterprises (MNEs) resident in foreign tax jurisdictions. The inherent lack of domestic capital in most developing countries has also resulted in most of their mining projects being funded through debt and equity facilities provided by the financial subsidiaries of their foreign holding companies generally resident in low-tax jurisdictions involving cross-border remittance of relevant interest and dividend payments. As it will be discussed in more detail later in Chap. 9, determination of appropriate levels of borrowing and interest rates relating to these often complex funding transactions is extremely complex and has created opportunities for tax minimisation or even avoidance in the country hosting the mining operations.

As a reaction tax authorities in developing countries have reacted by including in their tax laws provision for withholding tax to be levied on the remittance of certain types of payments abroad as a first line of defence. Withholding tax of 10–20% typically applies to the remittance of interest payments for loans sourced abroad, dividends to foreign shareholders, and remuneration for the services of consultants and other specialised personnel provided by foreign entities not taxed in the country. That is to say not through a local fixed place of business, a so-called 'permanent establishment' (PE).

Potential for possible abuse arises when a mining development becomes dependent on the availability of foreign loans as the overseas holding company may:

- borrow the whole or the majority of the necessary development capital abroad justifying the unusually high levels of debt because of the difficulty in raising adequate levels of equity or its excessive cost given the perceived country risk; and/or
- borrow the funds abroad at a low rate of interest on account of its high credit rating and on-lend them to the project subsidiary at a much higher rate of interest quoting once again country risk as the justification.

Most tax jurisdictions ward against excessive levels of debt by introducing within their tax laws 'thin capitalisation' rules limiting the maximum allowable level of debt to equity ratio in a project or taxable entity. These are frequently of the order of 2:1–3:1 (i.e. 66.67–75% debt). These ratios are well in excess of the average rate of borrowing observed in the mining sector in general given the illiquidity of its assets. Interest payments relating to borrowing in excess of these limits are reclassified as dividends and, as a consequence, are not deductible from revenue for the purpose of assessing taxable income.

Notwithstanding as thin capitalisation provisions are not mining specific, they in practice act as a fiscal incentive for mining because, under normal conditions, average mining debt to equity ratios tend to be much lower, generally less than 50%.

The situation is more complex when it comes to determine what an appropriate arms-length rate of interest should be. The problem is generally addressed in one of two ways by tax authorities either capping:

- the interest rate applicable to foreign loans in the country; or
- more recently, as recommended by the OECD BEPS Action 4, the total amount of interest deductions claimable by the local subsidiary as a specified benchmark proportion of the earnings before interest, tax, depreciation and amortisation (EBITDA) of the taxable entity (fixed ratio rule) or of the consolidated group of companies whichever the highest (group ratio rule).

Partial or full exemption from withholding taxes may however be provided:

- to foreign entities resident in jurisdictions with which the country has entered into a bilateral or multilateral double taxation agreement (DTA); or
- by means of special agreements concerning individual specific projects.

The first instance is a strong motivator for MNEs in locating some of their service-providing subsidiaries in countries that have DTAs with the country hosting the mining project. This is a tax-minimisation process often referred to as 'treaty shopping.'

Many tax jurisdictions levy a tax on the value of fringe benefits provided by the mining companies to their employees. Of particular significance is the value of the accommodation provided free or at a subsidised rent to employees in remote mining towns and in some case the value of meals and other amenities. Although this tax should be imposed on the employees, sometimes the mining company would pay it on their behalf. The tax represents a strong disincentive for mining companies to build mining towns and or support the development of long-term accommodation by unrelated parties in remote areas and encourages commuting arrangements for their workforce such as fly-in fly-out (FIFO) rosters. Reductions or exemption from fringe benefit tax (FBT) is therefore commonly provided by government to encourage broader regional development through the establishment of towns and related facilities.

Take Away Points

Government's point of view

• To the extent that a perception of prospectivity is the drawcard in attracting foreign direct investment (FDI), fiscal incentives directed to promote exploration tend to be most effective. These include collection and free dissemination of regional geoscientific information, exploration investment incentive (EIS) schemes, such as 50% subsidies to drilling in greenfield projects, and favourable tax treatment of gains on disposal of exploration tenements.

• Flow through shares appear to have been effective in stimulating investment in junior to mid-size exploration companies in Canada and to a lesser degree in Australia.

• Accelerated depreciation (at the limit immediate deductibility) of mining-specific assets is the fiscal incentive of choice in most jurisdictions. By deferring taxation accelerated capital recovery allowances encourage investment in new assets and increase cash flows in the early and most vulnerable years of mining projects.

• By contrast depletion allowances are undesirable as they do not encourage investment in new assets and are hard to administer.

• R & D concessions while desirable need to be based on a clear definition and mutual understanding of what constitutes R & D.

• Tax holidays are generally undesirable because, if based on an unconstrained time period (e.g. 3–5 years), may encourage abuse in the form of high-grading or increases in production rates both of which reduce the residual life during which the project will pay taxes.

• Tax holidays should be limited to CIT and not include mineral royalties, which are not a tax but a payment for the minerals belonging to the community. They should be constrained in terms of the tonnages of ore or metal that would be allowed to be extracted during the holiday period in line with the project feasibility study.

• Tax holidays should be based on a tax rate of zero percent, i.e. depreciation should be deducted and not carried forward as a loss to be deducted once the holiday expires in effect adding further years to it.

• A clear definition is needed as to what constitutes the project and how to handle subsequent satellite discoveries.

• Government should ensure that adequate cash is available for VAT and other tax refunds when due and avoid unwarranted delays which have been the source of significant taxpayers' discontent.

Mining industry's point of view

• Industry should be fully conversant with current incentives and strongly lobby for adequate funding for their chosen country's Geological Survey and for the introduction of various exploration incentive schemes.

• It is in its interest to foster and contribute to a system whereby the annual exploration reports are factually accurate and comprehensive and in a format that facilitates their incorporation into the country's geoscience database facilitating their early release.

• Industry should structure itself to derive early benefits from immediate write-offs and/or accelerated depreciation if possible by offsetting them against the taxable income of an associated company. This may not be possible if the project is held and operated by a company which is ring-fenced for tax purposes.

• Tax holidays would be desirable and should be sought for. However, to avoid serious disputes it would pay to structure them unambiguously on the basis of the tonnages of ore or metal which were forecast when submitting the project feasibility study to government and/or advising and getting approval by government on a timely fashion of any unforeseen changes to them.

• Clear definitions should be framed and agreed with government as to what constitute the project and how to handle future discoveries on the relevant leases.

• Industry should attempt to negotiate exemptions rather than refunds for some of the imposts such as VAT, custom duties and excises so as not to be subject to potential delays.

References

Abimanyu A (2016) Dilemma of Tax Incentives for Investment in Asia-Pacific: Indonesian Case. Paper presented at the 13th Annual Asia-Pacific Tax Forum: Meeting the Challenges of the Future, Jakarta

Babajide N, Ogunlade C, Aremu D, Oladimeji S, Akinyele O (2014) Comparative analysis of upstream petroleum fiscal systems of three petroleum exporting countries: Indonesia, Nigeria and Malaysia. Int J Sci: Basic Appl Res 15(2):99–115

Desai D, Jarvis M (2012) Governance and accountability in extractive industries: theory and practice at the World Bank. J Energy Nat Resour Law 30(2):101–128

Easson A (2001a) Tax incentives for foreign direct investment Part I: recent trends and countertrends. Bull Int Tax 55(7):266–274

Easson A (2001b) Tax incentives for foreign direct investment Part II: design considerations. Bull Int Tax 55(8):365–375

Easson A, Zolt EM (2002) Tax incentives. World Bank Institute, Washington, DC

Guj P, Martin S, Maybee B, Gosai N, Cawood F, Bocoum B, Huibregtse S (2017) Transfer pricing mining with focus on Africa: a guide book for practitioners. World Bank, the IM4DC and the CET, Washington, DC

Holland D, Vann RJ (1998) Income tax incentives for investment. In: Thuronyi V (ed) Chapter 23, Tax law design and drafting, vol 2. International Monetary Fund, Washington, DC

Kraal D (2019) Petroleum industry tax incentives and energy policy implications: a comparison between Australia, Malaysia, Indonesia and Papua New Guinea. Energy Policy 126:212–222

Luo D, Yan N (2010) Assessment of fiscal terms of international petroleum contracts. Pet Explor Dev 37(6):756–762

Mining Tax Incentives Database (2019). https://www.bing.com/search?q=mining+tax+incentives+database+2019&form=EDNTHT&mkt=en-au&httpsmsn=1&msnews=1&plvar=0&refig=123ca62dfade4766b6dbfdceeb33e9e1&sp=-1&ghc=1&pq=mining+tax+ince&sc=0-15&qs=n&sk=&cvid=123ca62dfade4766b6dbfdceeb33e9e1

Readhead A (2018) Tax incentives in mining: minimising risks to revenue. The International Institute for Sustainable Development and the Organisation for Economic Co-operation and Development

Zolt EM (2015) Tax incentives: protecting the tax base. In: Trepelkov A, Tonino H, Halka D (eds) United Nations Handbook on Selected Issues in Protecting the Tax Base of Developing Countries. United Nations, New York, pp 451–495

Chapter 6
Quantitative Financial Analysis of Impacts of Mineral Royalties on Project Economics and Resources Sterilisation: A Case Study

Abstract While governments strive to secure income from mining operations through taxes, royalties and other fiscal measures, it should not set these imposts to a level that discourages foreign direct investment (FDI) or that sterilises an excessive amount of mineral resources or even worse renders projects uneconomic thus inhibiting their development. The most common imposts affecting a mining project viability, as discussed in this chapter, are the specific weight-based and the ad valorem mineral royalties.

Keywords Project economics · Royalties · Economic impacts · Sterilisation

6.1 Description of a Gold Mine Case Study and Analytical Methodology

This section aims to provide, by means of a realistic example, analytical processes that will assist a:

- Government in defending a royalty on non-renewable natural resource exploitation; and
- A mining company in creating an argument as to why the imposition of excessively high royalties may prove to be a disincentive to investment.

The analyses will make use of and provide numerical evidence derived from a Case Study (Appendix 4) involving the discounted cash flow (DCF) financial model of an operating gold mine with specific parameters as outlined in Table 6.1.

The Base Case DCF model was interrogated to analyse changes in project value, taxation, welfare costs and life of mine (LoM), under different royalty rates. The impacts that the changing parameters have on the Base Case have been numerically presented in tables or in graphs.

Additional operating and economic parameters are assumed for the Base Case example:

E. Lilford, P. Guj, *Mining Taxation*, Modern Approaches in Solid Earth Sciences 18,
https://doi.org/10.1007/978-3-030-49821-4_6

Table 6.1 Input parameters for Base Case DCF model

Geological resource	Mt	32.00
Diluted mining reserves	Mt	30.09
Cut-off grade	Au g/t	2.99
Annual ore production	Mt	0.30
Recovery	%	91.0%
Total operating cost	$/t milled	122.80
Final rehabilitation cost	$mill	30.00
Total capital expenditure	$mill	325.00
Sustaining capital	% of capex	4.0%

- The gold price applied to determine the economic ore cut-off grade: A$1750/oz. (US$1320/oz., or A$54.431/g);
- The US$:A$ exchange rate: 0.754: 1;
- Ore Reserve (tonnes above the cut-off): 30.09 Mt (Mt is million metric tonnes);
- Milling capacity of 3 Mtpa (million tonnes per annum).

The ore production output feeding to the processing plant is constrained by a fixed milling capacity (Cairns 1998), in this case being 3Mtpa. On this basis, the cut-off grade (CoG), which is the lowest grade at which ore can be mined to generate break-even pre-tax cash flows (excluding the impact of capital expenditure as this may be sourced through equity or debt), was determined to be 2.99 g/t gold. Based on the above collective assumptions, the parameters would be sufficient to cover an estimated total operating unit cash cost (OpCost) of A$122.80/t milled for the operation, as illustrated below:

Total operating cost (A$/t mined/milled) $=$ OpCost $= P \times$ CoG
Where:
$P =$ gold price (A$/g of recoverable gold contained); and
CoG $=$ cut-off grade (g/t Au), which includes the impact of mining dilution and mining losses.

In order to break even, the LoM plan may include ore production from various sources within the mining operation blended to achieve an ore grade delivered to the mill of approximately 2.99 g/t.

Due to the variable nature of mineral grades within a mineralized body, trade-offs between tonnages and grades economically exploitable at different levels of commodity prices (the higher the commodity price, the greater the economic tonnage incorporated into the LoM plan) are possible. These trade-offs are often represented as a grade-tonnage curve in graphical or tabulated form. Figure 6.1 below shows the grade curve for the case study. The calculation of the curve assumes that no royalty is levied against revenue and, since the operation would not make a profit at the CoG, income tax is also excluded from the calculations.

It is important to note that under the exact economic parameters (gold price, exchange rate, production volume and operating costs used to calculate the cut-off grade) for the ore as delivered to the mill at the exact cut-off grade, the operation should pay zero corporate income tax because its taxable income would be zero

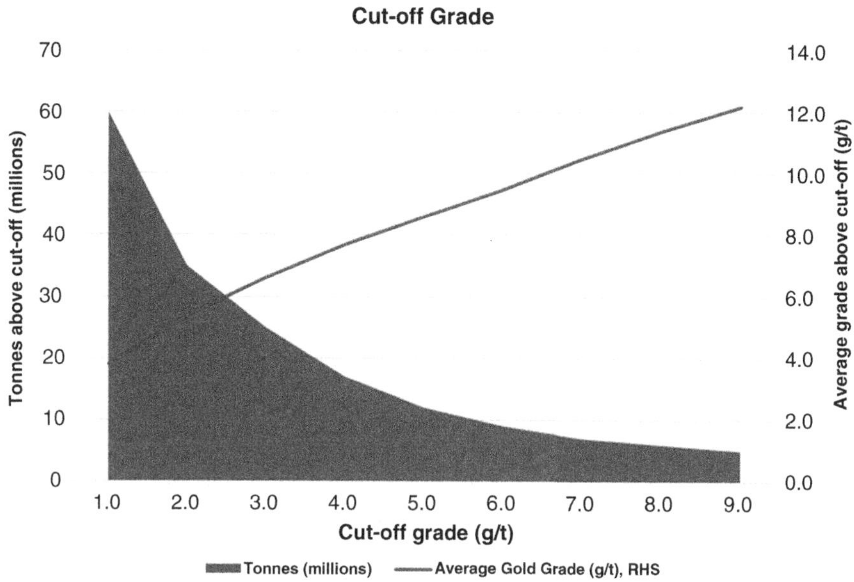

Fig. 6.1 Assumed cut-off grade curve—Case Study—Gold Mine

(absolute break-even). This is only achievable academically since it is not likely that each of these parameters would in practice comply simultaneously with the assumptions used, notably over a number of consecutive periods.

Since specific or unit-based royalties as well as value based or *ad valorem* royalties are charged on the income of an operation, and assuming the royalty rate is known, then the project for which the cut-off grade curve is being determined will need to include the effect of this royalty in the calculation. This is because this royalty may be deemed as an additional pre-tax or operating cost. In effect, the ore body will have to be mined at a higher average grade in order to break even after one of these royalties is applied without realizing a profit. Naturally, since the investors in the project want a return on their invested capital, the targeted grades modelled in the LoM plan will need to be higher than the originally-determined cut-off grade. That is, a higher cut-off grade will be calculated due to a specific or *ad valorem* royalty.

From the analysis completed in this Chapter, Table 6.2 below provides a summary of the results determined from the Case Study under a 3.0% and a 7.0% income royalty scenario. The supporting details and complete analysis provide below an explanation of how these results were derived and what they mean.

Table 6.2 Summary values arising from the analysis

Scenario		Base case	3.0% royalty	7.0% royalty
Cut-off Grade	g/t	2.99	3.08	3.20
LoM reserves	Mt	30.09	28.34	26.39
Break-even operating cost	A$/t	122.80	126.48	131.40
Saleable gold sterilized	kg	–	2688	5552
Value sterilized	A$'mill	–	146.29	302.20
State's royalty income	**A$'mill**	–	**107.54**	**242.69**

6.2 Impacts of a Specific or *ad valorem* Royalty at a Rate of 3.0%

In the gold example outlined previously (highlighted in Appendix 4), consider the case in which the State levies a 3.0% specific (weight-based) or an *ad valorem* royalty against the income generated by the operation. Since the gold mine is operating under the exact parameters outlined above, the operating, pre-tax cash flow calculations are performed at the cut-off grade. On this basis, it is reasonable to assume that for every one percentage point variation, up or down, in the operating revenue, being linearly correlated to the domestic gold price, the impact on pre-tax cash flows would be the same as if one varied the total operating cost by the same one percentage point, up or down. This is not a perfect assumption for the same reasons outlined previously and as described in Lilford (2003), but is used to highlight the impact that the imposition of such a royalty has on a mining operation.

From the above outlined assumption, the 3.0% *ad valorem* royalty in this example is determined to have a similar impact on the operation as would increasing the total operating costs by 3.0%. These parameters are deemed to be linearly correlated.

Therefore, increasing the total operating costs by 3.0% takes the pre-determined costs from A$122.80/t (A$41.057/g) milled to A$126.48/t milled, which equates to a total operating cost of A$42.288/g of gold. The impact that this increase in cost (or reduction in revenue) would have on the operational performance of the project, with all other parameters remaining unchanged, would be to move the project into a loss-making position while still operating at its pre-determined cut-off grade. This would likely not occur in practice because management would need to obviate this negative impact, and therefore the operation would need to mine higher grade ore at the same mining rate previously achieved in order to move back to at least a break-even cash flow position.

Calculating the impact, the cut-off grade will need to be increased by 0.090 g/t to maintain the same break-even cash flow position compared with the case where no specific or *ad valorem* royalty is considered. The cut-off grade has increased from 2.99 g/t to 3.08 g/t. Taking the revised cut-off grade onto the grade tonnage curve of Fig. 6.1 it can be observed that the available economic tonnage (Reserve) to be included in a revised LoM plan has reduced by approximately 1.75Mt from 30.09Mt to 28.34Mt of ore at a revised cut-off grade of 3.08 g/t.

Stated differently, the imposition of a 3.0% specific or *ad valorem* royalty in this example has sterilized 1.75Mt of ore. This royalty has reduced the economic component of the orebody such that the original LoM plan calculated the economic results of the operation based on a total of 89,996 kg of gold being mined, while the 3.0% *ad valorem* royalty case will economically exploit 87,308 kg of gold over the new mine life. An amount of 2688 kg of gold has been sterilized due to the imposition of this royalty. If the total recovery of ore to payable gold bullion was taken to be 80%, being a somewhat conservative assumption, then the loss of potentially saleable gold is 2150 kg over the life of the mining operation, which at current market values, implies that the value sterilized due to the royalty is approximately A$146.29 million. This is not a discounted cash flow value but rather a calculated point value of the total amount of gold that would be sterilized at the assumed economic parameters provided above, and assuming that the production could take place upfront (all at once) and not periodically over time.

Similarly if the State were to receive its total 3.0% *ad valorem* royalty as a lump sum payment (i.e. not a discounted cash flow value of the royalty) based upon the total production over the mine's life, it would receive 3.0% of total 87,308 kg of gold-equivalent value in the form of a royalty. Using the same economic assumptions provided previously, the State's receipts due to the 3.0% *ad valorem* royalty, and assuming the same 80% gold recovery rate, would be A$107.54 million on an undiscounted basis.

While the sterilized value of A$146.29 million and the State's royalty income of A$107.54 million differ by almost A$40 million, a greater issue lies in the interpretation of what the figures actually imply. This will be discussed later in this section once a similar analysis has been outlined for an *ad valorem* royalty of 7%.

6.3 Impacts of a Specific or an *ad valorem* Royalty at a Rate of 7.0%

Considering another scenario, being that of a specific or an *ad valorem* royalty rate of 7.0% of revenue, the impacts on ore sterilization will intuitively be significantly greater, as will the royalty receipts to the government's treasury. The Case Study gold project as outlined previously will be used again to demonstrate these impacts, which can then be compared to the 3.0% *ad valorem* royalty rate and zero royalty rate scenarios as summarized in Table 6.2 above.

Recalculating the cut-off grade will follow the same process discussed under the 3.0% *ad valorem* royalty rate scenario. The overriding assumption is that the 7.0% *ad valorem* royalty rate applied to the project revenue from gold sales will have the effect of increasing the total operating cost by a similar amount of 7.0%. This effectively increases the total operating cost by 7.0% from the Base Case amount of A$122.80/t milled (A$41.057/g) to A$131.40/t milled (A$43.931/g of gold). Therefore the cut-off grade will need to increase to an amount that achieves the same break-even cash flow (zero profit and zero loss) as was achieved in the zero royalty rate scenario.

The revised cut-off grade under this scenario increases to 3.20 g/t, being 0.209 g/t higher than the Base Case (zero royalty) cut-off grade of 2.99 g/t. From the grade curve in Fig. 6.1, the total economically mineable amount of ore to be modelled in the LoM plan (Reserve) is now approximately 26.39Mt, 3.70Mt less than the original 30.09Mt Reserve. This 3.70Mt reduction is the amount of ore sterilized due to the *ad valorem* royalty increasing total costs and consequently increasing the cut-off grade for a similar economic benefit in the calculation of Reserves in the LoM plan.

The 7.0% *ad valorem* royalty has sterilized 3.70Mt of ore and has consequently reduced the economically viable component of the orebody by this tonnage. The original mine plan modelled an amount of 89,996 kg of gold being mined and, at an assumed 80% total recovery rate, produced 71,997 kg of gold over its economic life. The 7.0% royalty rate scenario will economically mine 84,444 kg of gold producing, at the same 80% overall recovery rate, 67,555 kg of gold over the revised LoM plan. A total amount of 5552 kg of in-situ gold has been sterilized in this scenario. The in-situ sterilization value at a 7.0% royalty rate is calculated to be A$377.75 million with the recoverable gold sterilization value being A$302.20 million (80% gold recovery rate). Again, this is not a discounted cashflow value. It is a calculated point value, at the assumed price and exchange rate that was put forward in this paper, of the total amount of recoverable gold that has been permanently sterilized due to the *ad valorem* royalty of 7.0%. It again assumes that the gold production can take place all at once and not periodically over time.

In comparison, using the same economic assumptions and that the State would receive its 7.0% ad valorem royalty payment as a lump sum amount, the State would receive 7.0% of 84,444 kg of gold-revenue value derived from the sale of the gold in the form of its royalty. Therefore the State's receipts due to the 7.0% royalty would total A$242.69 million on an undiscounted, all-at-once basis.

Therefore, the 7.0% *ad valorem* royalty will sterilize A$302.20 million worth of gold while the State will receive A$242.69 million.

6.3.1 Royalty Rate Ranges

The impact of royalty rates of between 3.0% and 7.0% were tabled in the examples above because these are the rates most commonly imposed or discussed in most developing countries. Rates below and above the extremes of this range do occur in some jurisdictions and, for this reason, the impacts of specific and *ad valorem* royalties at differing rates has been determined and graphically presented in Fig. 6.2. In addition, and as discussed in Lilford (2017) and summarized below, the figure also shows the amounts lost due to lost salaries arising from employees

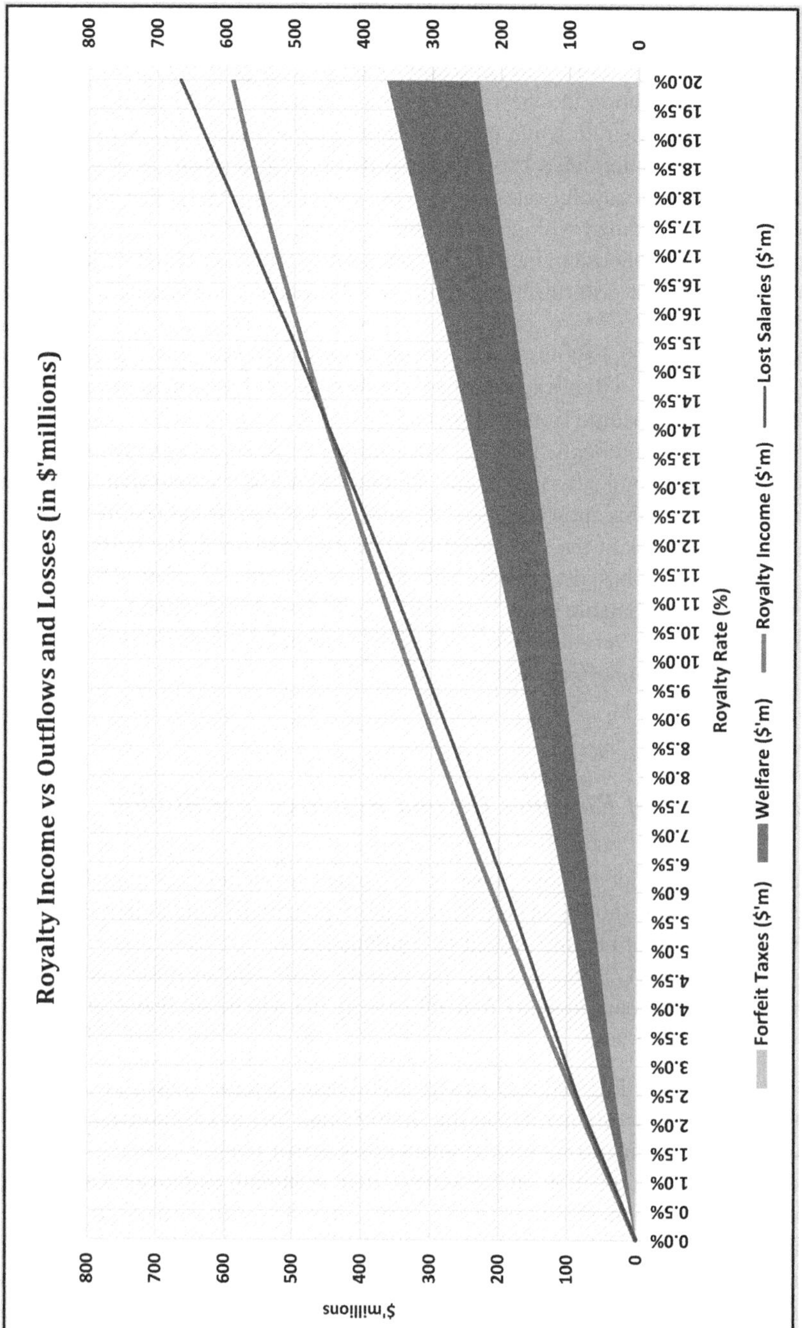

Fig. 6.2 State Costs, Lost Salaries and State Royalty Income for Varying Royalty Rates

retrenchments, their payroll taxes and the PAYE taxes forfeited resulting from a shorter mine life and employees being made redundant.

Figure 6.2 provides an image of the relative amounts of the various costs (losses and gains by the State and by the operation) due to a range of ad valorem royalty rates. The graph supports the view that the mine operator should ideally negotiate hard for a zero royalty rate while the State should argue otherwise. A compromise must inevitably be entertained between these two positions when, as is sometimes the case, ad valorem royalty rates, or at least exemption or payment moratorium periods, can be negotiated with government.

However, Fig. 6.2 does not include the comparative value permanently lost by the operation due to reserve sterilization. Figure 6.3 shows this comparison for different royalty rates.

The graph of Fig. 6.3 shows that a specific or *ad valorem* royalty, under current economic conditions, will always sterilize a higher value of gold than the State may possibly receive in income from that royalty.

This is because, although linked to price movements, an *ad valorem* royalty is still somewhat economically inefficient, although to a lesser extent than a specific royalty. In spite of this drawback specific and *ad valorem* royalties are the most commonly used types in the vast majority of jurisdictions, with the more economically efficient profit-based type only adopted in a minority of cases. This state of affairs is largely attributable to the latter being more complex to administer, easier to avoid, and generating very unstable flows of revenue and, in some cases, no revenue at all in the early years of production.

6.3.2 Impacts of Royalties Beyond Resource Sterilisation

The discussion provided above must be expanded to include additional royalty-induced impacts that should be considered before potentially concluding that an *ad valorem* royalty should be introduced or continued.

From Lilford (2017), the fact that the cut-off grade will increase under an *ad valorem* royalty and consequently mineral-bearing ore will be sterilized, also means the mining operation will:

- shorten its mining life (reduced Reserves and hence shorter LoM);
- retrench employees sooner due to the shorter mine life, discontinuing salary and payroll taxes and incurring retrenchment costs at that point;
- create a welfare imposition on the State sooner, assuming employees cannot find alternative employment;
- negatively impact any proximate community;
- discontinue utilizing services from various service providers earlier, who will in turn stop paying taxes arising from service payments from that operation; and

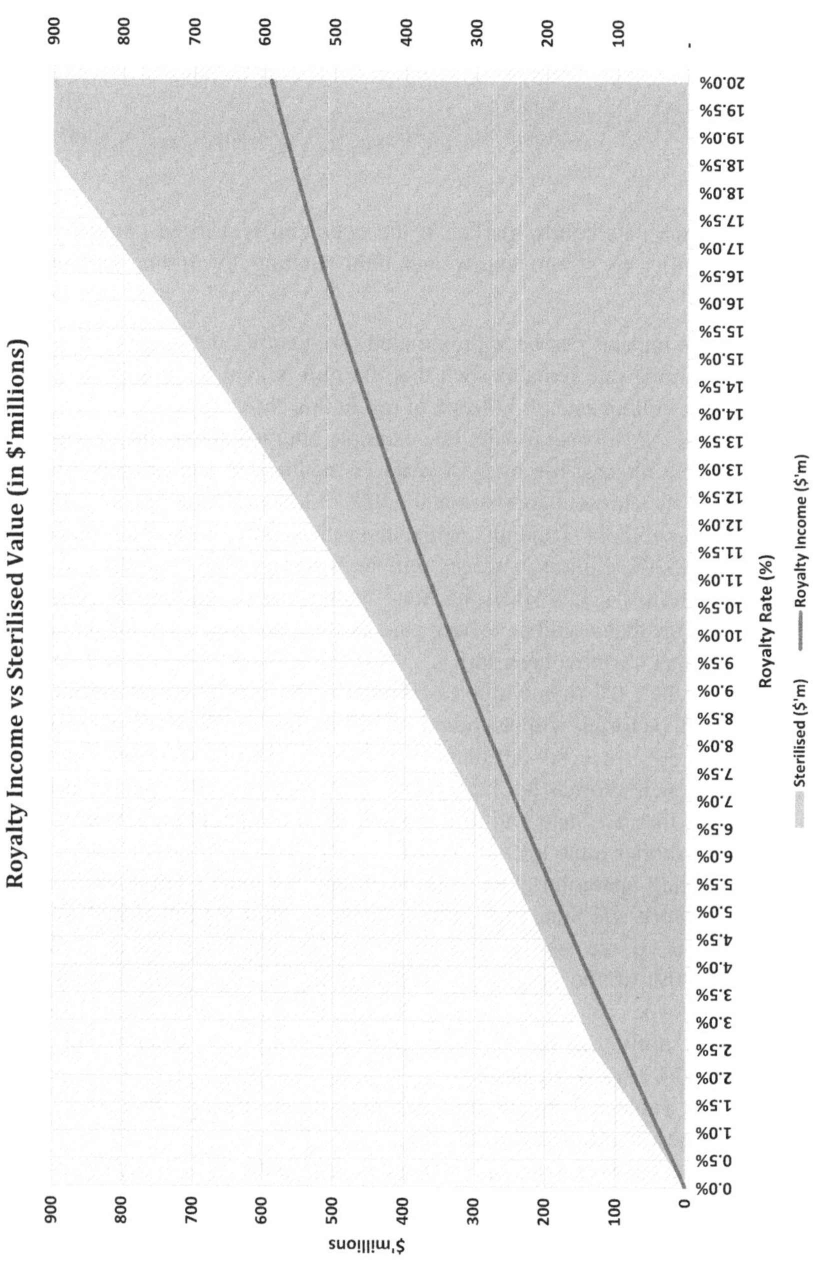

Fig. 6.3 Comparison Between State Royalty Income and Sterilised Value (in A$'million)

Table 6.3 Comparative financial positions (RR is the Ad Valorem Royalty Rate)

		Base case	3.0% RR	7.0% RR
LoM	Years	10.0	9.4	8.8
Royalty to state	A$'mill	0	108	243
Sterilized value	A$'mill	0	(146)	(302)
Opportunity loss (tax)	A$'mill	0	(38)	(80)
Welfare cost	A$'mill	0	(21)	(45)

- negatively impact its equity market, if the operation is a listed company, and reduce its ability to secure equity and debt funding for future growth and development.

Some of these impacts can be approximated. To quantify the impacts, it will be taken that the mining rate remains such that the mill is kept running at its design capacity, being a milling rate of 3.0Mtpa of ore in this case.

From the 3.0% *ad valorem* royalty rate example outlined above, the royalty will shorten the mine's forecast life by 0.58 years (7 months). If it is assumed that the mine's labour costs represent approximately 50% of the total mining costs, then by ceasing mining operations 7 months earlier than the original mine plan proposed, approximately A$107 million in wages will be forfeited. If we take the average employee tax rate to be 35%, then the State would lose around A$38 million in personal income tax that would have been paid by the employees over the remaining 7 month period that has now been lost.

Similarly, for the 7.0% royalty rate case scenario, the operating costs attributable to labour that will be forfeit will be around A$227 million (3.7Mt of ore sterilized at A$131.40/t and 50% of costs attributable to labour). The LoM under this 7.0% royalty scenario will shorten by 1.23 years. Therefore the 35% income tax rate assumed implies that the State will lose around A$80 million in employee income taxes due to the shorter mine life.

Once the mining operation closes, and assuming the mine's employees cannot find alternative work, the State may become liable to offer the retrenched workers welfare payments. If we assume that these welfare payments are at a rate of approximately 20% of the employee's previous salary rates, then an additional 20% of the employee cost will translate to an additional cost to the State. This will amount to A$21 million in the 3.0% *ad valorem* royalty rate scenario and A$45 million in the 7.0% royalty scenario.

In summary, Table 6.3 below provides the various figures discussed above (calculations appear in Appendix 4).

While the State receives a notable financial benefit in the form of royalty income from the operation, it will permanently lose some income from personal taxes (redundant employees) and will probably have to pay out welfare support amounts to some of those redundant employees. The royalty income to the State is higher than the lost taxes combined with welfare payouts, so the State may argue that this is acceptable. However, the State will typically not recognize the sterilization value because, in conjunction with the above financial impacts, it does not support the argument against the introduction of a royalty.

Operators often enter into commodity hedges in order to minimize the risk of incurring losses during a ramp-up period or a period in which it is expected that commodity prices (or exchange rates) will negatively impact cashflows. In addition, if an operator borrowed money in the form of debt from a financial institution, the financial institution may require that the operator hedges a certain volume of its future production. The amount hedged will typically match, with an added margin for costs and profit, the outstanding debt. In the situations where the operator has hedged some of its production, should the State charge its *ad valorem* royalty against the actual operator's revenue receipts (sales income incorporating the impacts of hedging) or should the *ad valorem* royalty be charged against the income that would have been derived under the commodity's spot price, ignoring the hedge positions?

A similar question that can be asked is where the operator does not produce a finished product, being a metal in a typical resources case. If the operator produces a matte or a concentrate or any other unrefined product, it will be required to pay treatment charges and/or refining charges or further beneficiating charges to a third party in order to produce the final, saleable metal. Will the State recognize this and allow for deductions against revenue for further processing and refining costs (these are cost and not revenue items) or will it charge its *ad valorem* royalty against the derived revenue and ignore these additional costs? The implications either way are significant and should be analyzed on a case-by-case basis.

6.4 General Principles

A number of papers and publications have been drafted providing research and commentary around royalties in the resources sector (Lilford 2017). One preferred publication that was supported by The World Bank is by Otto et al. (2006) titled Mining Royalties. This publication provides details and worked examples to provide the reader with a clearer understanding of royalties within the mining sector. The publication details taxation in the minerals sector, mining royalty instruments, the impacts of a royalty on a mine, investors and society at large, as well as governance and managing the benefits derived from a royalty.

However, there is a notable limitation to the publication in that its Chap. 4 discusses and quantifies the impacts of royalties on three worked examples of different mines. The three different mines reflect mines hosting different commodities, being copper, gold and bauxite. Otto et al. (2006) delve into quantifying how a royalty impacts the life of a mine by reducing the economic tonnage of ore available as a royalty is introduced and that royalty rate increases, based on a sensitivity. The limitation is that the worked examples and the commensurate explanation do not explicitly address the fact that the cut-off grade varies as a royalty varies. The worked examples simplistically reduce the mines' lives by inflating operating costs until negative cashflows are realized, and then subtracting the non-royalty, total mined tonnage by the revised royalty-reduced tonnage under the royalty scenarios. Nevertheless, the publication is a recommended read for further understanding of the issues.

Take Away Points

Government's point of view

• Since natural resources are national assets, exploitation of these natural resources must provide benefits to the country. Other than employment, skills and many other social benefits, governments wish to ensure that they receive a fair share of the economic benefits of exploitation through taxes, royalties and other fiscal measures.

• *Ad valorem* or unit-based royalties tend to be the preferred royalty regimes imposed in many emerging markets because they are simpler to manage and monitor.

• To attract foreign investment into their countries' mining sectors, transparent and clearly defined royalty rates are imposed on the sales of mineral products derived from mining activities.

• The potential impacts arising from these royalties realising a shorter mine life, being displacement (redundancies) of employees sooner and likely welfare costs after mine closure, are not adequately considered by Governments. There is an economic comparative to be considered when determining an appropriate royalty rate for the mining sector or specific commodities within that sector.

• Governments wish to levy as high a royalty rate as possible against mining production while attempting to not discourage investment into the sector.

Mining industry's point of view

• Royalties are deemed to be just an additional tax, negatively impacting the operation's cash flows and returns.

• *Ad valorem* and unit-based royalties result in mining operations targeting higher cut-off grades to ensure that the same returns would be achieved in the event that no royalties were levied at all.

• Increasing cut-off grades implies that the final Reserves and Resources figures are reduced and what was once marginally profitable ore that may have been blended with higher grade ore through mining has now been permanently sterilised due to the royalty.

• Companies want to pay as low a royalty rate as possible so that their operations can generate as high a return as possible over as long a period of time as economically possible.

References

Cairns RD (1998) Are mineral deposits valuable? A reconciliation of theory and practice. Resour Policy 24:1–10

Lilford EV (2003) The effect of the proposed royalty bill on mineral property values. J South Afr Inst Min Metall 103(6):337–344

Lilford EV (2017) Quantitative impacts of royalties on mineral projects. Resour Policy 53:369–377

Otto J, Andrews C, Cawood F, Doggett M, Guj P, Stermole F, Stermole J, Tilton J (2006) Mining royalties – a global study of their impact on investors, Government and civil society. The World Bank, Washington, DC

Chapter 7
Government Participation and Domestic Equity Requirements

Abstract Some countries, and notably emerging market countries, tend to either retain an equity stake in the mining operation on a non-payment, non-contribution basis, or legislate that local participation is required in the operation at an equity level. Local participation tends to be on commercial terms. In addition to these so-called "free-carried" interests, the same governments may also have an option to acquire an additional equity stake, at fair market value, in that operation.

Keywords Government equity · Local equity · Participation · Free carry

7.1 Participation Principles

In addition to taxes and royalties, and to ensure that Governments retain a deemed active involvement in and receive an appropriate financial benefit from their State's natural resources through mining activities, some countries have introduced a free carry, otherwise known as a non-participative or non-contributory, equity requirement for mining and minerals projects. Participative or carried interests are less popular, but may also exist either in addition to the free carry or as an alternative to a free carry.

A free carried, free carry (FC) or non-participative (non-contributory) interest (NPI), with each term being synonymous, is an equity interest at the local holding-company, project or asset level. This equity stake effectively grants the State, or a government subsidiary body, a predetermined specific equity interest in the operating asset. If the holding is FC or non-participative, this equity stake effectively exempts the holder, i.e. the Government, from contributing to an acquisition (entry) cost, any capital costs required to construct and develop the resources project or to expand or recapitalize it. The FC interest simultaneously grants the State or its subsidiary the right to receive its *pro rata* share of profits generated by that project once it is in operation. Payments to Government may be realised in the form of dividends based as either a percentage of the profit-after-tax generated by the operation or of its distributable income. Of notable importance, the percentage of the State's FC interest cannot be diluted. This implies that if the company decides to

E. Lilford, P. Guj, *Mining Taxation*, Modern Approaches in Solid Earth Sciences 18,
https://doi.org/10.1007/978-3-030-49821-4_7

raise additional funds by issuing new equity at either asset or holding company level, it is compelled to increase the Government's holding on a pro rata basis to ensure it is not diluted, at no cost to them.

The economic attributes of non-contributory and non-dilutionary FC interests will be explained and the implications considered in greater detail in this section.

It is important to point out that FC interests are only held in development or operating projects and not in earlier stage prospecting or exploration projects. This may be a moot point since the Government body holding the FC interest will not contribute to the exploration costs anyway. In fact, even if the equity interest was not FC, but participative or contributory in nature, the relevant Government body will not contribute to exploration anyway. This may create a perception by the investors in an exploration company that it holds rights to 100% of the project, which is initially true, but which will not be the case beyond the construction and development phase.

While a royalty is invariably levied on mining activities in most jurisdictions, a non-participative and non-dilutionary interest may also be imposed. In effect royalties and FC interests are not mutually exclusive arrangements and hence either one of the two may be applied separately, together or not at all.

Many emerging market countries impose a free carry right, in addition to a royalty, including:

- **Ghana**—10% FC interest, but unlike many, the 10% is only applied to dividends (Owusu et al., 2016). Royalty of 5% of gross revenue (gold);
- **Guinea**—15% FC. Royalty between 1.5 and 5%;
- **Senegal**—10% FC. Royalty typically between 3 and 5% (gold);
- **Tanzania**—16% FC. Inspection fee of 1% on export sales and Royalty averaging 6% (gold);
- **Mali**—10% preferred shares*. Typical Royalty of 6% (gold);
- **South Africa**—Mining Charter III proposes 10% FC (5% FC by employees and 5% FC by community). Royalty rate between 0.5 and 7%;
- **Burkina Faso**—10% preferred shares*. Royalty between 3 and 5% (gold); and
- **Democratic Republic of the Congo**—10% FC. Royalties of 3.5% on gold and base metals and 10% on strategic metals including cobalt-.

*Preferred shares are shares that provide the holder preferential rights over any other shareholder and, in the cases above, provides the State with the right to receive distributions (dividends) irrespective of whether other shareholders receive any distributions.

7.2 Free Carried Interest

Predominantly imposed by an emerging market's government, a number of African and a few other non-African countries have adopted the requirement that the State itself, or an appointed State subsidiary entity, must hold a specific equity stake in the

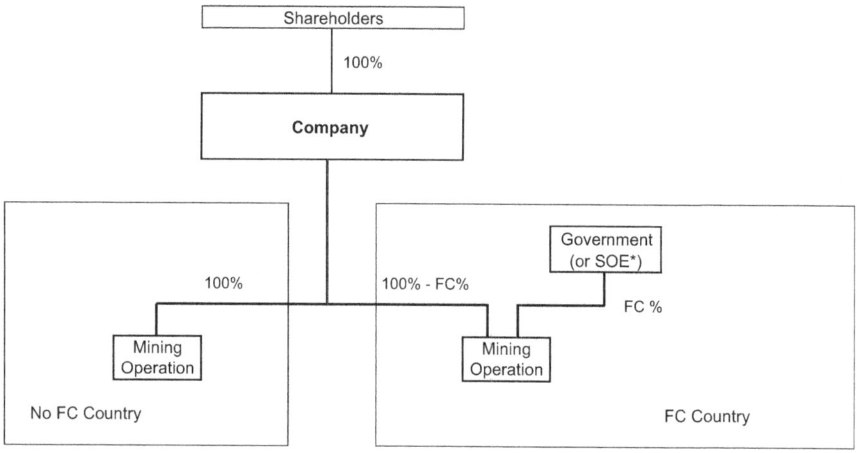

Fig. 7.1 Typical corporate structure including and excluding FC

local operating resources project or in the local project-holding company. In addition, a few of the same countries also allow for the government to acquire, at fair value, an additional stake in the project or local holding company.

The term "free carry" or "free carried" interest has been used historically to reflect this non-contributory equity holding but the term is being met with ever-increasing resistance. While use of the term is frowned upon and has actually diminished, what it represents and its continued regulated imposition remains as steadfast as ever. In this section, the terms free carry/ied (FC) interest and non-contributory or non-participative interest (NPI) will be used interchangeably, as they imply the same outcome.

A FC or NPI means that the holder of the -interest has a direct participation in the local equity of the operating project and hence the benefits arising from the investment are similar to any other shareholder in that entity. Generally, the FC interest is held in the local subsidiary company owning the asset or directly in the asset and not in the holding company which is often a non-local, listed (or unlisted) offshore company, as shown in Fig. 7.1. The State's FC equity stake is legally not saleable, but corrupt disposal practices cannot be totally excluded. In reality it would be very difficult to actually sell a FC equity holding in an operating mining subsidiary, as it is generally a legal entity not listed on any stock exchange. That it, the shares making up the FC interest have zero liquidity (no market for trade) and hence there are almost no buyers for that stake. Only a strategic investor would purchase such a stake.

As stated, the FC equity holding differs from other shareholders' positions in that the former is non-contributory in nature, as well as being non-dilutionary.

Being non-contributory means that the NPI holder firstly, does not purchase its stake for any consideration and secondly does not need to provide any capital to the project at any stage of its development or operation. So when the local company or asset spends development, expansion or recapitalization capital (collectively referred to as Development Capital) on its project, the holder of the FC interest is not required to contribute to the Development Capital requirements on a pro-rata basis based on its shareholding.

By way of example, if the State holds a 10% FC interest in a project and that project requires Development Capital of, say, $300 million, the State will provide no funding at all and the remaining shareholders, typically the owners of the holding company, will have to fund the full $300 million required. In a 10% carried funding or participative scenario, the State would be required to contribute its pro-rata share, that is to say $30 million, of the required Development Capital amount, with the other 90% shareholder(s) having to fund its collective pro-rata share of $270 million.

Since the FC interest is non-dilutionary, after the funding or the Development Capital contribution, the State will still hold 10% of the equity in the project and the holding company or other shareholder(s) will still hold 90%. A FC interest or NPI cannot be diluted by any financing (equity or debt), cash-call or other equity or pseudo equity funding or incentivising event. In addition, the FC interest generally cannot be sold or transferred to third parties, unless that third party is a wholly-owned State entity. Therefore, in most cases, the State or its subsidiary cannot divest of its FC interest.

Additionally, despite not contributing to the development cost requirements, the FC interest cannot be diluted through further capital calls. The inevitable consequential impacts of these arrangements on the economic attractiveness and other aspects of the operation will also be discussed in more detail later in this chapter.

It is also important to note that the FC equity interest held by the State at the local level cannot be rolled up (converted, exchanged or sold) into the offshore holding company's equity base, which may be listed on a public exchange. This is generally the case to ensure that the Government cannot exit its FC holding through the sale of its shares in the market, as this would breach the local FC rules and theoretically, in turn, force the local company to issue additional FC shares to the Government, thus diluting the remaining shareholders further. This may be a risk that could potentially arise under a change in Government or while under the governance of a corrupt Government (van der Ploeg 2011).

This situation finds a parallel in the context of Government's requirements for the establishment of mandated local equity levels in mining projects as for instance in the case of the Black Economic Empowerment (BEE) Charter of the Republic of South Africa discussed below.

Under the auspices of the Mineral and Petroleum Resources Development Act (MPRDA 2002), the South African charter rules contained in the Black Economic Empowerment (BEE) Charter of the Republic of South Africa (Chamber of Mines of South Africa 2003) do not stipulate a government-held FC interest in minerals projects, but the above issue around selling of regulated BEE interests arises. The current position, which was contested before reaching resolution by numerous parties, is that in the event that the BEE party sells any of its equity holding to a non-BEE third party, the affected company will not be required to issue new equity to another BEE party to maintain its regulated 26% BEE status. The implication is that "once empowered, always empowered" will prevail. 2018 changes to the charter stipulated that the 26% Broad-Based BEE (B-BBEE) participation is to be increased to 30%, with at least 7% of that amount being held by the workforce (with 5% of that 7% to be on a FC basis) and another 7% to be held by the local community. Therefore only 16% will be held by B-BBEE holders outside of the community and the operation's labour.

Due to governing law, a FC interest cannot actually be owned or taken by the State for no consideration. Consequently there has been, as mentioned earlier, a move away from this terminology. To get around the law, the NPI is formally secured or justified on the basis that the issue of a mining license by the State carries the necessary value to effect the exchange (acquisition) for the non-participative equity portion. The State typically owns the minerals while the mining license grants to the operator the right to extract and sell the minerals subject to requirements for the payment of leasing fees, taxes, possibly royalties, social and environmental commitments and other obligations. Effectively, therefore, the operator sells, and the State acquires the FCI with the sales price being the mining license itself.

The NPI stake has no financial commitment or outflow obligations on the side of the Government in securing its equity. It does, supposedly, grant the State significant benefits, which it could be argued mathematically provides the State with an infinite rate of return on its equity exposure and investment.

A number of the current NPI agreements grant the State the right to receive dividends from the operation when that operation pays a dividend. Most new mining operations, however, do not pay dividends for the first few years after commencing operations because they prioritize the generation and availability of free cashflows for the repayment of debt and/or to provide adequate initial equity returns to shareholders through further capitalisation of the project. It is generally only once the financial gearing of the operation has been reduced to a desired ideally sustainable level that dividends may start to be paid out. This means that the State's NPI may only provide income to its treasury after many years of holding the equity. However, the State does not have to contribute to the upfront or ongoing capital expenditure outlays, which are often significant amounts. One of the issues here is that it's likely the government personnel will have changed over the same period of time, so no credit will be forthcoming to the Ministers-of-the-day when the FC was first entered into. This unfortunately seems to be more important that the benefits that will ultimately accrue to the country as a whole later on.

Other than receiving dividends, some NPI agreements incorporate preference shares in the operation or local company that grant the State the right to receive a preferential financial benefit arising from the operation irrespective of whether a dividend is paid or not. This is significantly more punitive on the operation and its non-State shareholders than the case where the State only has the right to receive normal dividends, if paid.

Most of the countries that have legislated a NPI obligation also have regulations that allow the State to acquire an additional interest above the NPI in an asset or operation. The additional acquisition of equity above the NPI must be conducted on an agreed-value basis, with the value generally being determined by an independent third party. Naturally, in emerging markets, there is seldom any funding available to acquire additional equity and typically the State asks the operator to advance to it a shareholder's loan to support the acquisition. This raises the question as to whether a shareholder's loan for the acquisition of an additional equity stake in a project should be priced at the cost of debt, that is to say at typical debt funding rates, rather than at a rate equivalent to or higher than the cost of equity (Lilford 2006; Lilford et al. 2018).

Similar to the ability to acquire an additional stake, some countries legislate that local participants other than the Government must participate in the equity structure of the asset or local company. This may be in addition to the State's NPI, such as the case in the Democratic Republic of the Congo, or at the exclusion of a NPI as for example as envisaged by the Black Economic Empowerment Charter (Chamber of Mines of South Africa 2003). As is typically the case with the State having inadequate funding to pay for its additional equity stake up-front, local participants and communities generally have even less access to the requisite participation funding, so shareholder's loans have to be entertained in these situations too. The legislated equity participation of local participants is another significant issue requiring further attention and is discussed later in this section.

7.3 Impacts of a Non-Participative Interest

The State's legislated requirement for a NPI, not surprisingly, has a significant influence on the decision as to whether to invest in a mining project in that country at all. The fact that the holding company or local operation must provide 100% of the required Development Capital but will only receive 90% of any future dividends or any other financial benefit, assuming the State's NPI is 10%, means that the holding company's investment returns will be reduced.

In the case of the State's NPI benefit being based on dividends, from an economic position, investment returns are typically not calculated on dividends but rather on distributable income (profit after tax), since distributable income may be reinvested to further capitalise, expand or further develop an operation. It may therefore be argued that the operator could reinvest all of the profits generated back into recapitalising or expanding its operation and never pay out a dividend or only pay out a small dividend that is a low percentage of the available distributable profits. Realistically, this is highly unlikely since the non-Government shareholders of the operation, who provided the upfront Development Capital and any additional sustaining capital, will probably be expecting dividends at some time in the future rather than just capital (equity or share-price) growth.

The total project value, as defined by its net present value (NPV) which is derived from a post-tax discounted cash flow (DCF), is:

$$NPV = {}_{t=0}{}^{n}\sum C_f/(1+r)^t - I_c \tag{7.1}$$

where:

C_f = anticipated free cash flow at time t in real/nominal terms;
t = time/period in which the cash flow occurs;
r = real/nominal discount rate;
I_c = present value of the capital outlay to bring the project to account.

However, with an X% FC interest, the State's pre-tax attributable value will be:

$$\begin{aligned} PV_s &= X(NPV) + X(PV_{capex}) \\ &= X\left({}_{t=0}{}^{n}\sum C_f/(1+r)^t - I_c\right) + X\left({}_{t=0}{}^{n}\sum I_c/(1+r)^t\right) \end{aligned} \tag{7.2}$$

where:

PVs = present value (PV) attributable to the State, and the company or asset's attributable value will be:

$$\begin{aligned} PV_c &= (1-X)(NPV) - X(PV_{capex}) \\ &= (1-X)\left({}_{t=0}{}^{n}\sum C_f/(1+r)^t - I_c\right) - X\left({}_{t=0}{}^{n}\sum I_c/(1+r)^t\right) \end{aligned} \tag{7.3}$$

On a post-tax basis, these can be written as:

$$\begin{aligned} PV_s &= X(NPV) + X(PV_{capex}/(1-T)) \\ &= X\left({}_{t=0}{}^{n}\sum C_f/(1+r)^t - I_c\right) + X\left({}_{t=0}{}^{n}\sum I_c/((1+r)+(1+T))^t\right) \end{aligned} \tag{7.4}$$

where:

T = corporate tax rate,and the company or asset's attributable post-tax value will be:

$$\begin{aligned} PV_c &= (1-X)(NPV) - X(PV_{capex}/(1-T)) \\ &= (1-X)\left({}_{t=0}{}^{n}\sum C_f/(1+r)^t - I_c\right) - X\left({}_{t=0}{}^{n}\sum I_c/((1+r)+(1+T))^t\right) \end{aligned} \tag{7.5}$$

The example put forward in Chap. 6 has been used to discuss the impacts that a NPI has on the operation. Since a NPI is seldom introduced without an existing or introduced royalty, the discussion that follows reflects the combined positions of both an *ad valorem* royalty and a NPI.

Figure 7.2 provides the Base Case NPV at varying discount rates for the gold operation of the Case Study discussed in Chap. 6 with no royalties and no NPI imposed. The NPV of the operation is calculated on a 100% equity ownership basis, paying corporate income tax at a commonly applied 30% tax rate with no royalty.

The internal rate of return (IRR) of the operation in the Base Case is calculated to be 32.5%.

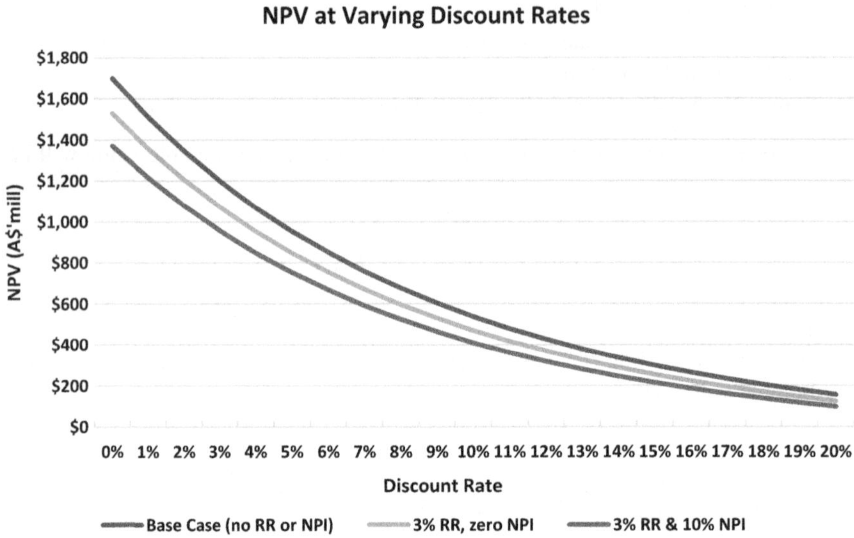

Fig. 7.2 Operation NPV at Varying Discount Rates, including a 3% Royalty and 10% NPI

In the same Fig. 7.2, a curve showing the impact that a 3% royalty will have on the NPV of the same project at varying discount rates is also shown. As discussed previously, the impact of a 3% *ad valorem* royalty will reduce the available economically exploitable tonnes of ore from the Base Case amount of 30Mt to 28.3Mt and increase the mined gold grade from the Base Case of 2.99 g/t to 3.07 g/t, both of which are factored into the curve. The third curve in Fig. 7.2 shows the impact that the 3% a*d valorem* royalty combined with a 10% NPI will have on the NPV of the operation.

The IRR of the operation paying a 3% royalty with no NPI is 30.3% while the IRR of the operation paying a 3% *ad valorem* royalty while simultaneously being subject to a 10% NPI is calculated to be 28.5%. The 10% FC reduces the return to ordinary shareholders by almost 2%.

Figure 7.3 shows the benefit to the State measured as a present value (PV) of the 3% royalty, of the 10% NPI and of these two imposts combined at varying discount rates. The IRR to the State is infinite since these benefits are achieved without the latter having to provide any investment capital into the project.

It is clear from Fig. 7.3 that the State has very little to lose under either one or both of a 3% royalty and/or 10% NPI while the operation's owners will need to consider whether the calculated returns and value of the operation justify its development (commitment of capital) in the first place. It is justifiable to argue that, since the State is not exposed to the same cost of capital and other risks to which the operation's owner is exposed, the discount rate for the State's exposure, its so-called social discount, should not be as high as that for the owner and operation itself. Perhaps the discount rate that the State should use should only reflect its sovereign risk as reflected by the yield on a typical Government bond. That is to say its discount

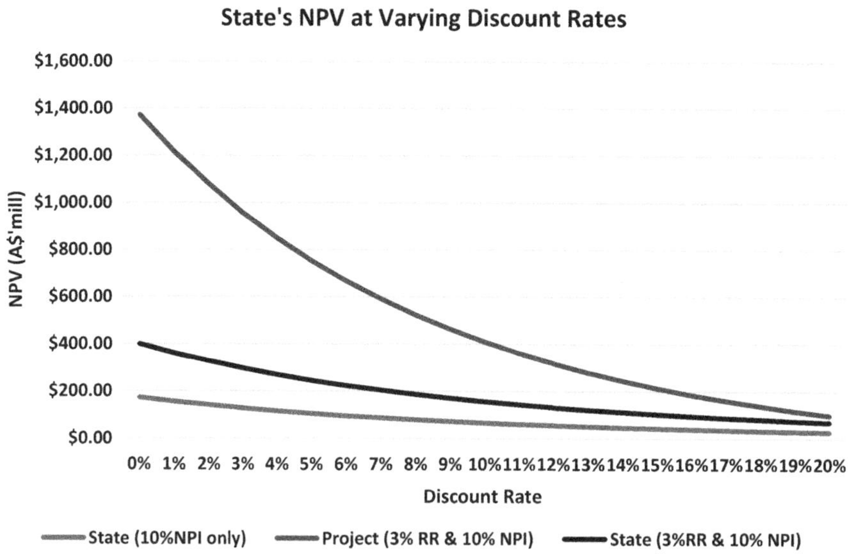

Fig. 7.3 PV of the State's benefits from a 10% NPI and a 3% Royalty Combined

rate should only be its risk free rate. Refer to Lilford (2018) for further discussion on this aspect covering the cost of capital and discount rates.

On a similar basis, consideration is given to a 7% royalty rate scenario. Again, as discussed previously, a 7% *ad valorem* royalty rate will reduce the economically mineable tonnes to 26.4Mt (from the Base Case of 30Mt), shortening the mine's life, and increasing the cut-off grade to 3.208 g/t from the Base Case cut-off grade of 2.99 g/t of gold.

The IRR of the operation paying a 7% royalty is 27.1% while the IRR for the same operation paying both a 7% royalty while being subject to a 15% NPI is calculated to be 24.5%. These represent significant reductions from the IRR of the operation in the Base Case at 32.5%. The respective and combined effects of a 7% royalty and 15% NPI is graphed below in Fig. 7.4 for varying discount rates.

Once again, the State will be more interested in the following Fig. 7.5 because it shows the NPV attributable to the State at varying discount rates of the 7% royalty and the 15% NPI scenarios combined. The State's benefit curve is compared with the project's NPV under the 7% royalty and 15% NPI scenario in the diagram.

From Fig. 7.5 it is evident that the State has much to gain under either one or both of a 7% royalty and 15% NPI while the operation's owners and hence shareholders will need to consider whether the value and returns of the operation attributable to them can be justified.

Iterating, it is arguable that the State should consider using its risk free rate as the discount rate and not the project's discount rate.

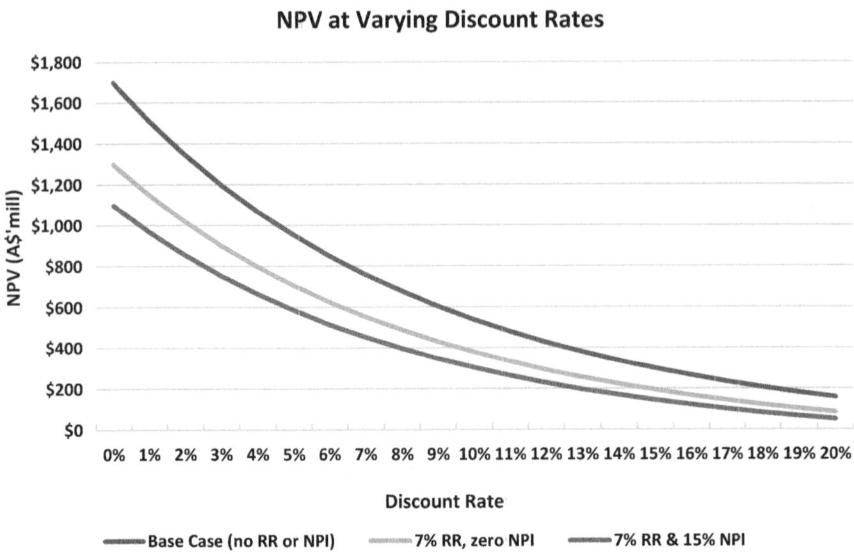

Fig. 7.4 Operation NPV at Varying Discount Rates for 7% Royalty and 15% NPI

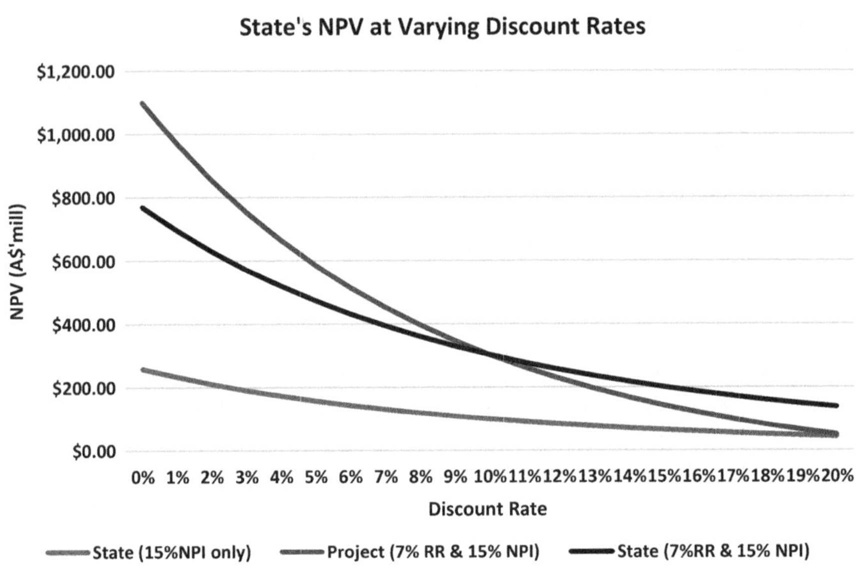

Fig. 7.5 NPV of the State's 15% NPI and 7% Royalty Combined

Table 7.1 Summary of Impacts Arising from a Royalty (3% and 7%) and a NPI (10% and 15%)

	Base Case	3% Royalty	3% RR & 10% NPI	7% Royalty	7% RR & 15% NPI
IRR Op	32.5%	30.3%	28.5%	27.1%	24.5%
NPV 8% Op	$677 m	$596 m	$525 m	$430 m	$397 m
NPV 10% Op	$538 m	$470 m	$411 m	$332 m	$303 m
NPV 12% Op	$427 m	$369 m	$319	$254 m	$228 m
NPV 10% State	**$0 m**	**$66 m**	**$157 m**	**$99 m**	**$304 m**

7.4 Summary Results and Analysis

In summary, the Table 7.1 below highlights the impacts of an *ad valorem* royalty and a NPI, individually and cumulatively, for selected real discount rates. The results tabled show the operation's value as a NPV and its IRR and compares these with the value attributable to the State under similar conditions. The results reflected for the State's income benefits exclude the income arising from corporate tax, which has been modelled at a rate of 30%. Therefore the operation values shown include the deduction of income tax assumed at a corporate tax rate of 30%.

It can be seen from the scenarios considered in this chapter that *ad valorem* royalties alone generate significant value to the State in the form of de-risked income while those same royalties simultaneously reduce the value and returns to the owners of the operation. If the State also introduces a NPI in addition to royalties, the combined impact must be considered with a significant focus on whether these impositions are such that they deter investment and hence the development of that country's resources opportunities. A balanced position must be put forward that adequately rewards the providers of the Development Capital to attract those investors, as well as to provide the State with sufficient benefits that can be shared with the nation.

The consequences of including a royalty and a NPI are significant to both parties. It is necessary for an operator to factor the inclusion of these economic inputs into his feasibility model to ensure that the operation at least meets the investment hurdle rate (often measured as an IRR) and demonstrates an adequate NPV at a predetermined discount rate. This assumes that the State has existing legislation prescribing these rates.

A significant risk to the operator arises when the State decides to change its policies and, once the operation is being developed or is already in production, introduces new legislation that demands a royalty and/or NPI when it had not previously.

It cannot be concluded that a royalty and a NPI are necessarily appropriate since they individually and collectively reduce value to the operation's shareholders while simultaneously increase the value and income to the State. It is the shareholders who provide the development risk capital in the first place, often with the support of bank loans and project financing, and therefore these shareholders will want an adequate return on their investment. Governments must also be compensated for leasing to the operator access to a non-renewable, national asset and hence must be paid taxes and possibly a royalty for this resource rental. However, a free carry or NPI is difficult to justify. Preferentially, if an equity participation is legislated, that participation should be held by the community/ies impacted most by the operation, or by the operation's employees, and not in the form of a free carry.

7.5 The (Un)Sustainability of Local Participation

From the discussion above, this implies that the local partner, either through a company structure or a community trust, must fund its own participation in the equity of the project. That is, it must acquire its equity interest and must also contribute to any future equity or other non-debt capital development costs the project undertakes. It is the securing of these funds that remains an issue, and thereafter, the ability to repay the funds at an equity-based cost of funding within a finite period of time.

Many emerging market countries including Zimbabwe, South Africa, Namibia, Tanzania, Zambia and Indonesia, have and continue to include and revise regulations that mandate foreign and local investors in the resources sector to structure and include the equity involvement of employees, local communities and/or domestic companies ("Groups") in their resources projects. By way of a specific example, revised legislation in South Africa proposes how future transactions and specifically Black Economic Empowerment (BEE 2003) transactions are to be contemplated. The original policy paper, Mining Charter 1, stated that 26% of any mining project was to be beneficially owned by a BEE Group. Through two further iterations, Mining Charter III has incorporated a revision legislating that proposes B-BBEE ownership be revised up to 30%. Of this 30%, 10% is on a free carried basis (5% to selected employees and 5% to the local community) with the remaining 20% to be held by BEE Groups.

The objectives of any local participation are typically twofold:

- It provides a local company or the local community nearest the project with an instrument that holds value that will potentially receive dividends from the project once and if it pays dividends.
- Secondly, the equity holding incentivises the holder to assist, where possible, with the provision of workers and community support in the development and operation of the project. The local participant may also provide support and a voice to the ears of government and other regulators that assist with the timeous awarding of various licenses and operating permissions.

It is important to quantify what level of compliance resources companies will have to meet to satisfy the requirements of the various participation policies with regard to ownership requirements for previously excluded Groups. In general, the funding structures available to these Groups may not necessarily provide a sustainable solution to their historical exclusion from the resources industry. The term sustainable in this context refers to the ability of that Group to convert its initially encumbered, or financially geared, equity holding into an unencumbered stake of equal size. Encumbrance in this context refers to the outstanding (unpaid) portion of borrowing, used to acquire the initial equity stake, that still has that equity stake pledged as security (encumbered) to the lender until the borrowed funds plus accruing interest have been fully repaid. Once fully repaid, the financiers' needs to hold any security is eliminated and therefore the equity holding is returned and is the holding is then deemed to be unencumbered.

The various legislations typically provide for the issue of ownership and participation to previously excluded Groups within the mining industry. The intention of the legislation is that ownership should be accompanied by effective participation to avoid "fronting". Fronting is a term used when a Group is granted an equity stake merely as window-dressing, with no meaningful benefit or inclusion offered to the Group. A meaningful benefit may include representation on a management committee or Board of Directors, receipt of distributions including dividends or other financial receipts, access to facilities, training, advancements and others.

The legislation also generally includes requirements for beneficiation, procurement, services, migrant labour development, women's participation in resources and other activities. Commitment by the company to a social and labour plan (SLP) is an important requirement of the various regulations and forms part of most company's sustainable development plan. Requirements of the SLP are generally very wide-ranging and the level of demand on company commitment is high. However, the administrative capacities of the relevant regulatory bodies to monitor and administer the practical outworking of the SLP is largely questionable.

In broad terms, the spirit in which the rules and regulations were written are defendable and supportable. They have been introduced in an attempt to avoid turmoil in the mining industry in their specific countries and to avert the possibility of nationalisation of natural resources projects. However increasing demands from certain quarters suggest that many failures have occurred in achieving these edicts. This can be seen with the creation of extreme wealth in certain instances to the exclusion of the majority of the country's populace (enrichment instead of empowerment) including an operation's labour and that operation's nearest local community.

Legislation has included more than equity and effective ownership, in that it is designed to bring an equitable and balanced approach to the beneficial ownership of mining interests to previously excluded participants. Consequently, this chapter describes at least one way in which resources transactions can be structured and what the implications may be due to the legislation.

7.6 Unencumbered Equity Ownership

7.6.1 Potential Implications to Achieve Unencumbered Equity Ownership

Generalising, it is hardly possible for a domestic company or the local community to access the necessary funding for the purchase of an equity stake in the operating asset. One of the main reasons for this is that financial institutions are extremely averse to proving debt-funding rates for a third-party equity investment. This then leads to the Group requesting financial assistance from the operator or owner on some repayable basis. However, even with financial assistance from the Company, it can be shown that the regulatory requirements result in an unsustainable equity participation in that Company's local mining assets. In addition, insistence on local participation often means that taxation on capital gains and taxation on dividends may arise resulting in value-leakage. Specifically value-leakage refers to the discounted value of the free cash flows dispensed or awarded to the taxation authorities in the guise of a secondary tax on companies (STC), potentially a capital gains tax (CGT) or some other taxation impediment. Thus value-leakage occurs as a direct consequence of a mining company being compliant with the relevant legislation.

7.6.2 The General Regulations

In general, the regulations will require that both locally and internationally owned companies engaged in mining activities must demonstrate ownership by local Groups to the extent of a minimum stipulated percentage holding within a finite period. Typically this ownership of local mining assets can be measured either in terms of equity participation or as a market share measured by attributable units of production. Specifically, Group ownership of mining assets may be stated and measured as:

- market share measured by units of production controlled by the Group;
- involvement in beneficiation activities or capacity for offsets against beneficiation allowing for some latitude;
- the continuing and anticipated-but-quantifiable future consequences of all previous deals would be included in calculating the credits/offsets in terms of market share as measured by attributable units of production; and
- where a company has achieved Group participation in excess of the target for a particular operation, then such excesses may be used to offset shortfalls in its other local operations.

In attaining the local equity participation target and excluding FC interests, Group participation will be established on a willing seller – willing buyer basis, at fair market value, where mining companies may be requested to facilitate this equity participation. This implies that mining companies may have to provide funding assistance in the form of shareholder's loans to the Group in order to be compliant with the regulations.

7.6.3 Quantifying a Group's Participation

7.6.3.1 The Objectives of the Regulations

The objective of the Group ownership of mining industry assets is to generally facilitate their sustainable entry into the mining industry and to allow them to benefit from the exploitation of mineral resources. Group ownership of mining assets is not considered a final solution, but rather a means of ensuring their access to the economic benefits of exploitation of natural resources. Consequently the final objective of the legislation is that the Group realises equity value from potentially the economically attractive extractive industries. This equity value inherent in mining assets can be estimated using market capitalisation or standard valuation approaches.

That is, the equity value of mining assets owned by a company listed on a public stock exchange is equivalent to the market capitalization of the company (share price multiplied by shares in issue). In the event that a company's shares are illiquid (not freely tradable), its market capitalisation may not necessarily reflect a fair trading value of the company. For non-listed companies, generally accepted valuation methods, such as the discounted cash flow (DCF) method incorporating a net present value (NPV) is most commonly used. An independent, third-party valuation specialist will be appointed to calculate and provide the valuation results.

7.6.3.2 Quantifying Group Participation

The two preferred methods of measuring a Group's ownership of mining assets are either equity participation or market share as measured by attributable units of production. In the event that Group ownership is measured by equity participation, the measurement must result in the realisation of equity value from the mining assets. Since all shareholders share in the asset's equity value *pro-rata* to their percentage shareholding, participation of a Group relative to that of other shareholders must reflect their proportionate exposure to the value of such mining asset.

In regard to actual extraction factors such as the current and forecast production profile of the asset, these may differ as the nature of the ore body unfolds during mining. What is estimated at a feasibility stage rarely stands up during actual mining. Differences in equity value and the revenues generated by production units is also a function of the selling price of the specific commodity produced. In classical mineral economics, commodity prices are subject to large and sometimes unpredictable

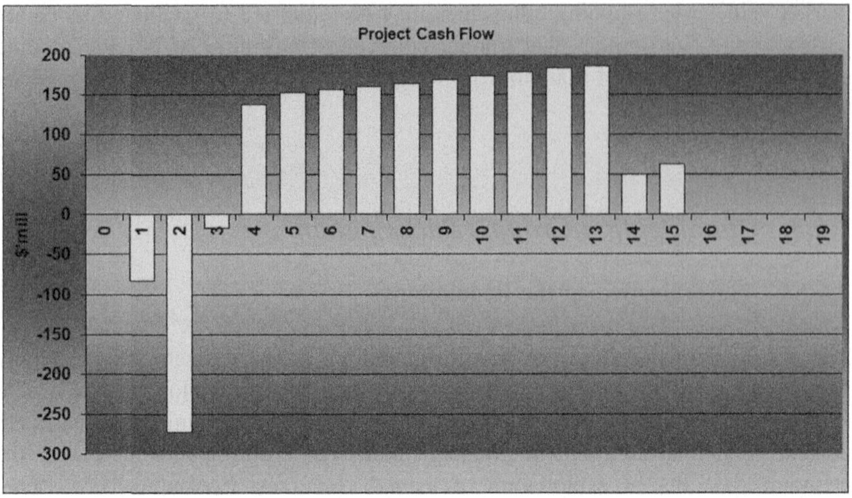

Fig. 7.6 100% of enterprise free cash flows

swings as factors affecting the global economy change and as events in the 2008–2009 global financial crisis period have shown. As commodity prices increase, the unit operating costs associated with the production of the commodity usually follows suit, albeit with lagged timing. Commensurately, only rarely will costs decline as commodity prices fall.

Significant differences in the equity value and the value determined by units of production may arise because capital expenditure to bring an ore body to production and the actual production profile may be quite different. In addition, royalty rates applicable to the revenue generated by the asset and taxation rates on profits both constitute a cost to the mining operation and these may vary depending on how the royalty is levied. The levels of debt to equity funding in respect of the asset may mean that the preferential claims of debt over equity create a significant difference between equity and production value. Finally the risks inherent in the technical characteristics (size, shape, depth, grade, tonnes), the economics of location, the cost of capital and technology inputs, and the risks associated with uncertain and unsuccessful exploration, long lead times, and political risks, may all contribute to the differences between equity and production values.

It is therefore possible that a specific asset representing 25% of attributable production units of the larger portfolio, selected for disposal to a Group, could have a significantly higher or lower percentage of the equity value of the portfolio. In other words, attributable tonnes of production from different mining assets will not necessarily realise the same equity value per tonne.

For illustration purposes, with reference to the determination of the equity value, Fig. 7.6 shows annual enterprise free cash flows from a mining project which will be attributable to a 100% stake in the company's local operations. The significant

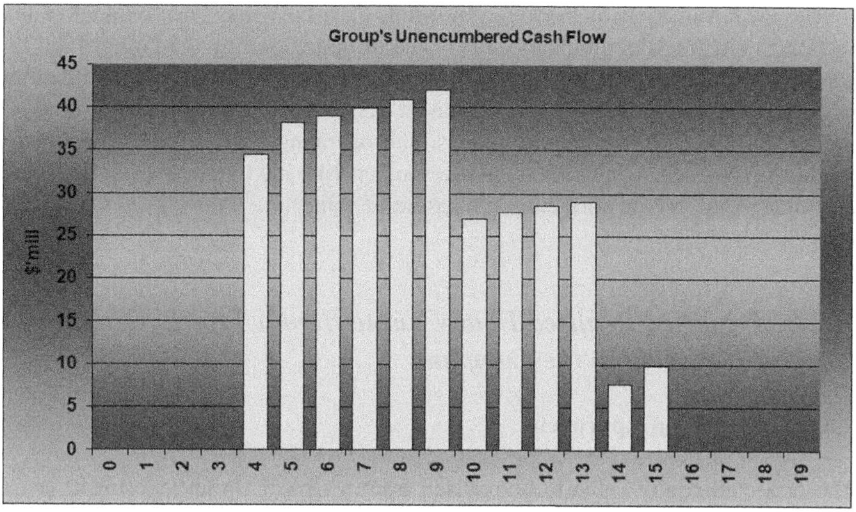

Fig. 7.7 Enterprise cash flows attributable to the Group from the original company at an acquisition price of 10% discount to fair value

negative flows in the early years represent the capital outlay required to bring the project into production.

Figure 7.7 shows the annual enterprise free cash flows attributable to the Group from an assumed transaction in respect of 25% of the equity of the company during and after the servicing of typical Group transaction funding. It was assumed in this example that the 25% equity stake was acquired at a price of 10% below fair value (i.e. a 10% discount price). Details of the calculation of the retained, unencumbered equity stake are covered in the next section. In general, however, over the first 9 years, all of the Group's attributable cash flows, being 25% of the total free cash flows less relevant taxes, will be used to repay its debt. Thereafter, the Group will hold an unencumbered 15.63% in the original company's local mining assets, as shown from Year 10 onwards in the figure, generating attributable free cash flows as indicated in Fig. 7.7.

7.6.3.3 Group Ownership Measured by Attributable Units of Production

Group ownership measured by attributable units of production would be expressed in terms of the expected equity value to be realised from production units. Two obvious advantages of this approach are firstly, it achieves alignment with the ultimate objective of ownership requirements, being the realisation of equity value by the Group from mining assets. Secondly, it results in equivalent measurement outcomes, whether the Group ownership is measured through equity participation or market share in terms of attributable units of production.

The DCF valuation methodology resulting in a NPV has been applied in this section to quantify the ultimate equity value to be realised by the Group from the control of attributable units of production. Consequently, all factors that affect the realisation of the Group's equity value in respect of such attributable units of production, including commodity prices, unit operating costs, capital expenditure, production profiles, royalty rates and taxes, other risks and third-party funding, have been integrated into an all-inclusive measure of value.

7.6.4 Projected Realised Equity Value from a Typical Group Investment in the Company

7.6.4.1 Valuation Approach

The projected equity value to be realised from a typical Group investment in the original mining company can be estimated using the previous example outlined above. It involves a Group transaction with a nine year duration in respect of 25% of the original company's local equity. The nine year duration is a typical debt-financing period for a mining project offered by financial institutions where a three year interest moratorium period is offered during which moratorium period interest is rolled up and capitalised. A valuation of 100% of the equity in the original company was performed to determine the current fair market value of a 25% equity interest in the company. Two typical company funding structures have been assumed, with funding assistance being provided by the company. Finally the likely value to be realised by the Group from the 25% equity investment in the company was determined, after repayment of the funding in respect of the previously mentioned Group funding structures.

7.6.4.2 Valuation of 100% of the Equity in the Original Company and Value of a 25% Interest

Based on the assumed cash flow valuation, the fair market value of 100% of the equity in the company is calculated to be $429 million. Therefore, the fair market value of a 25% interest in the company is one quarter of that, being $107 million.

7.6.4.3 Typical Group Funding Structures

As a result of a general lack of investment capital available for local communities or local companies with limited existing assets, historical precedent has proven that it is unlikely that a Group will be able to contribute a significant amount of its own financial resources to fund the acquisition of the 25% equity interest in the mining company. Therefore, the implementation of a Group transaction requires a funding structure whereby most, if not all, of the required funding will be raised from external sources. Typical funding structures have therefore been assumed in projecting the final Group equity value to be realised from a 25% equity interest in the company. Firstly sourcing and securing third-party funding by the Group for the acquisition of the 25%, where the original company will provide assistance to secure such funding by means of disposing of the 25% interest in the company at a discount to fair market value (Funding Structure 1). Secondly, funds may be secured by providing vendor funding on favourable terms to the Group by the company for the 25% acquisition, at fair market value (Funding Structure 2). Both of the previous funding structures suggest that the holding company will provide assistance to the Group for the acquisition through shareholders loans.

For the acquisition above, the Group is unlikely to secure funding from commercial banks. The reasons for suggesting this are that commercial banks prefer debt exposures to equity exposures, demonstrated by the fact that a significant portion of commercial banks' funding is in the form of debt. Equity exposures carry a higher and different risk to a debt exposure. Furthermore the company's future economic sustainability is dependent on the vagaries of the commodity markets. These markets are capricious as they are often impacted by global economic trends or other fundamental drivers. Consequently, the company bears additional risk associated with varying levels of supply and demand in the product market. Finally the company will be hard-pressed to produce significant free cash flows over the period for which commercial bank debt is normally provided, namely five to nine years, due to the inherent nature and associated margins of the mining industry in general.

Since the mandate of parastatal development institutions, such as the Industrial Development Corporation of South Africa Ltd. (IDC), the Public Investment Commissioners (PIC) and the Development Bank of Southern Africa Ltd. (DBSA), includes the funding of long-term Group exposures, it is likely that these and other similar institutions will be the source of third-party funding for many African and emerging market-based transactions. Since these institutions will bear the risk in funding the acquisition of the 25% Group interest, they will typically charge relatively high funding rates. The typical funding terms of parastatal institutions when funding long-term Group equity exposures include a base rate of 10–15% per annum in nominal terms after tax. In addition, a participation percentage in the proceeds realised from the underlying equity investment after redemption of the funding provided and the payment of all taxes, referred to as the upside, will also be called-for by the financier. This participation interest in the upside is generally around 50%.

In general, the steps in a Group transaction for the 25% equity investment funded by parastatal institutions will be to incorporate the Group interest in a special purpose vehicle (SPV) that will acquire the 25% in the company at a discount to fair value. The SPV then raises the full amount of the funding required for the acquisition through the issue of preference shares to the parastatal institutions. Assuming an upside participation percentage of 50% by the parastatal funding institutions, the SPV then issues 50% of its ordinary share capital to such institutions for a nominal consideration. This will create the mechanism through which the parastatal funding institutions participate in any potential upside (shown diagrammatically below).

The SPV simultaneously issues the remaining 50% of its ordinary share capital to the Group for a nominal consideration, to alleviate the funding burden on that Group. The ordinary shareholding of 50% in the SPV by the Group will enable it to participate in the upside from the company equity investment after repayment of the preference share funding. Any dividends received by the SPV in respect of the company equity investment will be, in the first instance, applied towards servicing the preference share funding provided by the parastatal institutions. Once the disposal by the SPV of the company equity interest has occurred (at some future date), the Group entity will realise value to the extent of 50% of the upside after redemption of the preference shares and the payment of all taxes.

In the event that the disposal proceeds relating to the equity investment in the company are less than the amount of the preference share funding outstanding to the parastatal institutions at the time, the Group will generally not be required to compensate the parastatal institutions for the shortfall.

7.6.4.4 Funding Structure 1: Projected Group Equity Value to be Realised

In Funding Structure 1, it is assumed that the original company disposes of a 25% equity interest in its assets to the Group for $96.5 million, which represents a 10% discount to its fair market value of $107.2 million, as determined through a DCF valuation ($107.2 million is 25% of the total value of $428.8 million). The Group will raise the entire funding requirement for the transaction from parastatal institutions on terms typically offered by these institutions when funding long-term equity exposures namely, a base funding rate of 13% per annum in nominal terms after tax plus a 50% participation percentage in the upside, and the company will re-acquire the Group's 25% equity interest at its fair market value after nine years, as determined through a DCF valuation at that point.

The third-party financiers are likely to bear substantial equity risk in funding the Group transaction, since the Group is unlikely to contribute a meaningful, if any, portion of the total funding requirement from its own resources and therefore will be unable to provide an equity buffer to the financiers. Furthermore the recourse of the third-party financiers will probably be limited to the value of the equity investment in the company, due to a lack of other financial resources that can be provided as security by the Group. The diagram in Fig. 7.8 illustrates the structure of the Group's equity participation in the company and the related funding structure, immediately after implementation of the assumed Group transaction.

The projected Group's equity value to be realised from a 25% investment in the original company was determined based on the assumption of an acquisition price of $96.5 million for its 25% equity interest. The issuing of preference shares in the amount of $96.5 million by the SPV to parastatal institutions is to fully fund the

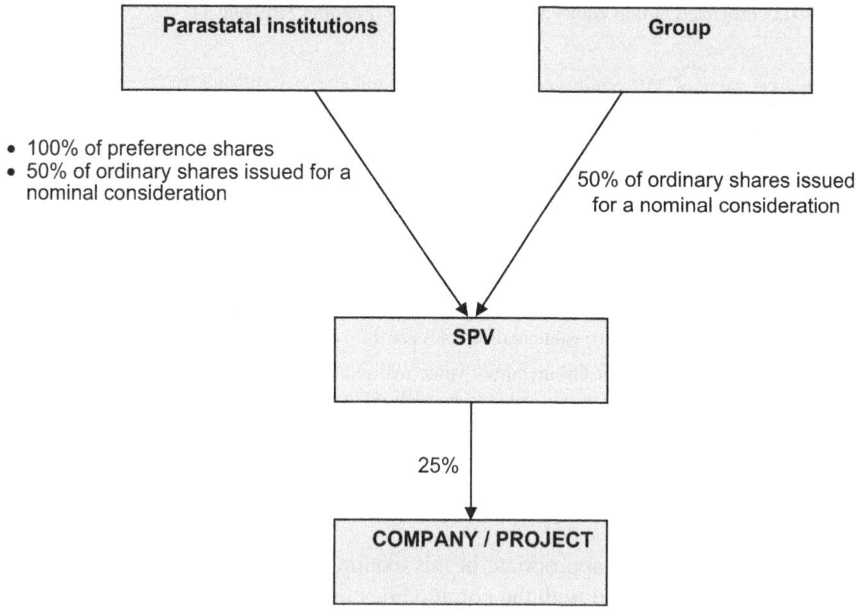

Fig. 7.8 Structure of the Group's equity participation in the company immediately after implementation of the transaction using Funding Structure 1

acquisition of the company's equity interest. Other than the 10% discount to fair value, the following funding terms apply:

- a base funding rate of 13% per annum in nominal, after-tax terms;
- a participation of 50% in the upside; and
- a transaction period of 9 years.

The disposal by the SPV to the Company, or another interested purchaser, of the equity interest at the end of Year 9 at its fair market value of $173 million at the time (undiscounted value in Year 0), where such value has been determined on the same basis and applying the same valuation assumptions in relation to the company as set out previously. This includes the payment of CGT or similar, if in a jurisdiction requiring this payment, by the SPV on the disposal profit in respect of the company equity interest, at an assumed CGT rate of 10%, the payment of STC by the SPV on net dividends paid at an assumed rate of 10%, and the application of the disposal proceeds in respect of the company equity investment after CGT and STC. These proceeds would go towards firstly, payment of the outstanding amount of the preference share funding, and secondly, towards payment of the upside participations by the parastatal institutions and the Group relating to their ordinary shareholdings of 50% each.

The projected Group value to be realised at the end of Year 9 was discounted at an appropriate risk-adjusted rate to a present value at Year 0. A risk-adjustment of 3%

Table 7.2 Projected Group equity value as at Year 0 (Funding Structure 1)

	$'m
Fair market value of 25% equity interest in the company at disposal (Year 9)	173
CGT payable on disposal profit	(8)
STC payable on dividend distributions	(10)
Disposal proceeds after CGT and STC	156
Outstanding amount of preference share funding	0
Total upside after repayment of preference share funding	156
Profit participation of parastatal institutions via their 50% ordinary shareholding in the SPV	(78)
Group equity value realised via its 50% ordinary shareholding in the SPV Year 9	78
Present value of Group equity value realised at Year 0	25

Note: For the present value of Group equity value realised at Year 0, a rate of 13.7% was used to discount the projected Group equity value to be realised at the end of Year 9 to a present value at Year 0. This is 3% higher than the discount rate used in respect of the cash flows of the company, due to increased technical, leverage and financial risk pertaining to the projected Group equity value to be realised at the end of Year 9

has been assumed to be appropriate in this example owing to the long-term anticipated volatility associated with the commodities industry in general.

The calculation of the projected Group equity value to be realised as at Year 0, assuming Funding Structure 1, is set out in Table 7.2 below.

The value indicated in Table 7.2 represents 5.72% of the total value of the company's assets. That is, after a typical funding term, and commensurate with associated funding conditions, the Group's original, geared, 25% equity participation in the Project would culminate in an unencumbered holding of 5.72% after the funding term of nine years. A shorter or longer funding term would impact on this unencumbered stake only slightly.

On a similar basis to that described above, if the financier structures repayment in which the financier does not take a 50% upside payment on transaction profits, then the Group will ultimately hold an unencumbered 11.45% in the original company's local mining assets, generating attributable free cash flows as indicated in Fig. 7.7 previously.

Each of the two scenarios presented above clearly shows that the Group will not be able to retain its initial 25% participation unless it can find its own acquisition funds. This is referred to as the unsustainability of a local Group transaction.

7.6.4.5 Funding Structure 2: Projected Group Equity Value to be Realised

It is assumed in Funding Structure 2 that the company and not a parastatal organisation will provide the Group with vendor funding to acquire the 25% equity interest in its local assets. The funding terms here are assumed as being more favourable than those available from typical third-party financiers. Specifically, the company will provide funding assistance to the Group by fulfilling the role of parastatal institutions such as those contemplated in Funding Structure 1.

Specific country regulations worldwide, such as Section 38 of South Africa's Companies Act, 1973, typically prohibit any company from giving direct or indirect financial assistance to others for the purpose of a purchase of or subscription for any shares of such company (or its holding company). Funding Structure 2 therefore needs to be proposed in a way that does not breach this general rule, which can add significantly to the transaction costs of a Group transaction.

The following funding terms have been assumed in respect of the vendor funding to be provided by the Project. This includes a base funding rate of 10% per annum in nominal terms after-tax, and no participation percentage in the upside on the equity investment in the company above the outstanding amount of the vendor funding. The vendor funding terms set out here are significantly more favourable than the typical funding terms of parastatal institutions when funding long-term Group equity exposures. This is so because the base funding rate of 10% is typically lower than the base funding rates of 10–15% offered by parastatal institutions, and the Project company will not participate in the upside.

7.6.4.6 Differences Between the Funding Structures

Whereas Funding Structure 1 assumes funding assistance by the company in the form of a disposal of the 25% equity interest at a 10% discount to fair market value, Funding Structure 2 assumes funding assistance by the company in the form of vendor funding on favourable terms to the Group, with no discount to the acquisition price. Consequently, it has been assumed in Funding Structure 2 that the company will dispose of the 25% equity interest at fair value.

In general, the assumed steps in a Group transaction relating to the Project which is vendor-funded by that company are to house the Group's equity interest in a SPV. The SPV will acquire the 25% in the Project taken at its fair market value and settle the acquisition price by issuing preference shares to the Project company for an equivalent amount. The SPV issues 100% of its ordinary share capital to the Group at a nominal value, thereby alleviating the funding burden on the Group. The ordinary shares in the SPV will enable the Group to participate in 100% of the upside from the equity interest of 25% in the company above the outstanding amount of the vendor funding. Any dividends received by the SPV in respect of the company equity investment will be applied, in the first instance, towards servicing the preference share funding provided by the company. The Group will realise upside to the extent that the value of the 25% equity interest in the company exceeds the outstanding amount of the vendor funding at the end of the transaction period. Finally the mechanism through which the Group will realise value is by means of a disposal of the SPV ordinary shares to the Project company at fair market value, where the fair market value of such ordinary shares will be determined as the fair market value of the SPV's underlying equity interest in the company less the

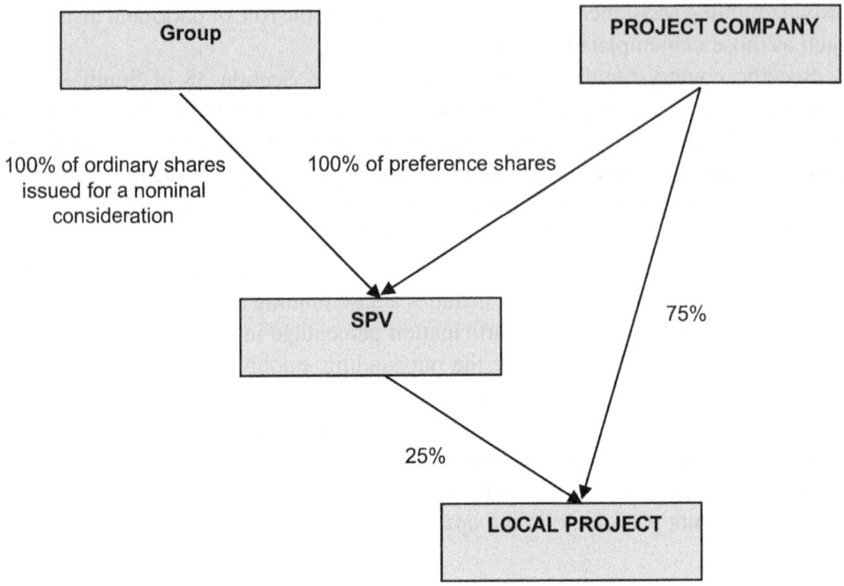

Fig. 7.9 Structure of the Group's equity participation in the company immediately after implementation of the assumed transaction using Funding Structure 2

outstanding amount of the preference shares issued by the SPV to the Project company.

In the event that the fair value of the SPV's underlying equity interest is lower than the outstanding amount of the preference share funding, the company will not have recourse to the Group for the shortfall. Consequently, the Project company will be assuming direct equity risk against the value of the SPV's interest in the company at a funding rate which does compensate it for such risk to the extent that the funding terms of parastatal institutions do.

The structure of the Group's equity participation in the Project and the related vendor funding structure immediately after implementation of the assumed Group transaction is shown schematically in Fig. 7.9.

The issuing of preference shares in the amount of $107 million by the SPV to the Project company, in exchange for the acquisition by the SPV of the company equity interest is based on typical funding terms. These funding terms include:

- a base funding rate of 10% per annum in nominal, after-tax terms;
- no profit participation in the upside on the company equity investment after full repayment of the preference funding; and
- a transaction period of nine years.

The disposal by the Group at the end of Year 9 of the ordinary shares in the SPV at fair market value of $173 million, where the fair market value of the SPV ordinary shares has been determined as the fair market value of the SPV's underlying equity interest in the company less the outstanding amount of their preference share funding (zero in this case as there is no outstanding funding after nine years). In turn, the fair market value of the SPV's underlying equity interest in the company has been determined on the same basis and applying the same valuation assumptions as set out previously, and the payment of CGT, if applicable, by the Group on the disposal profit in respect of the SPV ordinary shares, at an assumed 10%.

The projected Group equity value to be realised at the end of Year 9 was discounted at an appropriate risk-adjusted rate to a present value at Year 0. A risk-adjustment of 3% was again deemed to be appropriate.

The value calculated for Funding Structure 2 was conducted on a similar basis to Funding Structure 1 detailed above and represents 11.35% of the total value of the Project company's local assets. That is, after the funding term and the associated funding conditions as discussed above, the Group's original, geared, 25% equity participation in the Project would culminate in an unencumbered 11.35% in the Project after the funding term of nine years. This again demonstrates the "unsustainability" of the Group's equity participation over time.

In Funding Structure 2, it was assumed that the company will dispose of a 25% equity interest in the company to the Group at its fair market value of $173 million, as determined through a DCF valuation. It will provide vendor funding to the Group on terms that are significantly more favourable than the typical terms of parastatal institutions, namely, a base funding rate of 10% per annum in nominal terms after tax, no participation percentage in the upside, and acquire the Group's 25% equity interest in the company at its fair market value after nine years, as determined through a DCF valuation.

7.6.4.7 Value Leakage

The costs associated with implementing Funding Structures 1 and 2 are incorporated in the terms of the financing facilities. That is, other than some minor charges, the actual transaction implementation costs are factored into the interest rates and / or into the upside participation percentage. Additional "costs" that may be incurred due to the requirements of the relevant legislation include CGT charges on the calculated capital gain realised in the unwinding of the funding structures. This cost will be paid out of the free cash flows generated by the company.

In order for the Group to repay its debt, the company will have to issue dividends to the SPV. In some jurisdictions, including in South Africa, dividends may attract STC or some other form of dividend tax. Over the nine year financing term, a nominal amount of approximately $ten million covering the free cash flows will be payable as STC. At the same discount rate used on the cash flows, being 13.7% over nine years, the value at Year 0 of the STC is $1.5 million. This, together with the CGT (reflected in the preceding funding table) and the cost of funding represents the minimum value-leakage resulting from the required participation of a Group in the Project. The value-leakage arising due to STC applies equally to Funding Structures 1 and 2.

In addition, Funding Structure 1 demonstrates an additional value-leakage due to the equity value discount of 10% offered to the Group. Although Funding Structure 2 does not consider this discount, and the fact that the ultimate unencumbered stake held by the Group for Structure 2 is less than Structure 1, due to the significant facilitation by the Project company, it can be construed that this additional unencumbered equity participation also represents value-leakage to the company. That is, the difference in the unencumbered equity stakes once the funding structures unwind, being 5.63% of the company after nine years (11.35% less 5.72%), represents additional value-leakage over Funding Structure 2. This percentage leakage on the value of the equity amounts to an additional $2.74 million.

7.6.5 Summary Comment

Funding Structure 1 assumes funding assistance by the company to the Group in the form of a disposal of 25% of the equity in their assets at a 10% discount to fair market value. It is also assumed that the parastatal funding providers will charge 13% interest, after tax, and will participate to the extent of 50% of the upside. The unencumbered equity owned by the Group after nine years in this case is approximately 5.72% of the company's Project. In Funding Structure 1, the total value-leakage is calculated to be the sum of the discounted STC leakage ($1.5 million) and the discounted equity value offered to the Group ($10.7 million). That is, the total value-leakage in Funding Structure 1 is $12.2 million.

Funding Structure 2 assumes funding assistance by the company to the Group in the form of vendor funding for the full transaction price on terms that are significantly more favourable than the funding terms typically offered by parastatal institutions. The assumed after tax interest rate is 10% in this case. The unencumbered equity owned by the Group after nine years in this case is approximately 11.35% of the company's Project. In Funding Structure 2, the total value-leakage is calculated to be the sum of the STC leakage ($1.5 million) plus the increased unencumbered equity facilitation (cheaper funding by 3%, discounted) provided to the Group ($0.4 million). That is, the total value-leakage in Funding Structure 2 is $1.9 million.

This section demonstrates that the unencumbered Group equity value to be realised at Year 0 from a 25% equity investment in the company, assuming typical Group funding structures, will be substantially less than the initial transaction value before Group funding is taken into account. This is referred to as "unsustainable" participation.

The preceding analyses and discussion points highlight the financial imposition arising in large part due to STC and CGT. If these impositions were waived for Group transactions, the unencumbered equity stakes, after time, held by the Group would increase noticeably. Naturally, if funding costs were reduced or were subsidized by the Government, unencumbered stakes would increase further.

Take away points

Government's point of view

- By retaining a free carried or non-participative interest in a mining operation, Governments believe that they will receive income pro rata to their equity stake from the operation when it pays out dividends.
- A free carried interest provides a means for the Government's collective voice to be heard in the management of the operation.
- A typical free carried interest of 10% should not deter investment.
- Proximate communities or other local partners should also benefit from the mining industry and therefore some countries regulate that local communities or other local non-government equity participation is required. This should allow for skills-transfer and wealth-sharing.

Mining industry's point of view

- A free carried interest means that the Company will need to invest 100% of the capital development costs for a 90% benefit (if the free carried amount is 10%).
- Since investment returns are one of the key investment considerations in the first instance, Companies will factor in the free carried stake and target higher grade ores to ensure returns are not compromised.
- Companies manage their assets independently of Governments and therefore will not willingly be directed by the Government or any of its bodies, other than as required environmentally, socially, or by other regulations.
- There is a preference to re-invest capital for further equity growth rather than pay out dividends, partly due to the free carried interest and commensurate returns.
- Having local partners or including communities in the equity structure of the local operation is not the issue that it is perceived to be. The problem that does arise is that the local partners or community tend to have no access to funds, so the Company has to structure shareholder's loans to allow the former's participation. These loans are on commercial terms, but Companies question why they have to be a financier in addition to a mine operator.
- The free carried interest will never be granted at the holding company level due to the risk that the Government-of-the-day sells that stake and, when a new Government is installed, the Company will need to issue new equity to re-comply with the equity provisions.

References

BEE (2003). Broad-Based Black Economic Empowerment Act 53 of 2003, Assented to: 7 January 2004, Commencement date: 21 April 2004

Chamber of Mines of South Africa (2003) Memorandum to the national treasury on the draft mineral and petroleum royalty bill. Submission by the Chamber of Mines of South Africa, Johannesburg

Lilford EV (2006) The corporate cost of capital. J South Afr Inst Mining Metall 106:139–146

Lilford EV, Maybe B, Packey D (2018) Cost of capital and discount rates in cash flow valuations for resources projects. Resources Policy. https://doi.org/10.1016/j.resourpol.2018.09.08

MPRDA (2002). The Mineral and Petroleum Resources Development Act (No 28 of 2002). http://www.dmr.gov.za/Portals/0/mineraland_petroleum_resources_development_actmprda.pdf

Owusu O, Wireko I, Mensah K (2016) The performance of the mining sector in Ghana: a decomposition analysis of the relative contribution of price and output to revenue growth. Resources Policy 50:214–233

Van der Ploeg F (2011) Natural resources: curse or blessing? J Econ Literature 49:366–420

Chapter 8
Stabilisation Agreements

Abstract The investment capital typically required for any mining project is substantial. One of the key risks such an investment is exposed to is the possibility of a change in government once the investment has been made, where the new incoming government make unilateral changes to the fiscal rules (taxes, royalties, duties, participation, etc.) governing such an investment. To avoid or mitigate against such a risk, it may be possible for a company to enter into a stabilisation agreement with the relevant government which will stand enforced even in the event that the government changes.

Keywords Sovereign risk · Stabilisation · Security of tenure

8.1 General

Mine development can take place under two separate sets of conditions, being either:

- in compliance with the standard provisions of the prevailing mining legislation, e.g. **Mining Act and related Regulations**. Additional conditions specific to individual projects are frequently included when issuing the relevant mining titles; or
- under the terms of **Special Agreements** between government and industry. These may in turn be contracts negotiated between industry and a relevant government ministry or department (e.g. Mines or State Development), which may or may not then be ratified as an Act of Parliament in effect becoming law.

The Intergovernmental Forum on Mining, Minerals, Metals and Sustainable Development (IGF) has recently published a "Mining Tax Incentives Database" (2019) categorising the various types of fiscal incentives and detailing their application in 104 mining contracts across 21 countries. The fiscal incentives are categorised as those applying to mineral royalties, CIT and a host of other relevant imposts.

Although not required by law under most jurisdictions, investors often insist on special agreements to ensure that over the duration of their investment, the

E. Lilford, P. Guj, *Mining Taxation*, Modern Approaches in Solid Earth Sciences 18,
https://doi.org/10.1007/978-3-030-49821-4_8

regulations governing their equity exposure to a project do not change in any substantial manner. For this reason special agreements are often referred to as **Stability Agreements** as they ensure that the project owner's rights to the project cannot be taken away or be nationalised and that the conditions surrounding its economic exposure to the project are not compromised by changes of government or by unilateral government amendments to the regulatory and fiscal regime. From the project owner's point of view, the period of regulatory stability should ideally be over the life of the operation or possibly the life of the initial exploitation permit, which typically ranges between 20 and 30 years, with renewals available thereafter. It may also be over another defined period during which the project owner may have determined that investor returns will have been satisfied, although the government will not concern itself with investor returns.

In effect special agreements are used to establish a clear framework for the rights and obligations of the parties and for the ongoing relations between the State and a mining company tailored to facilitate the orderly development and operation of each individual major project. While some of the conditions may be common to a number of special agreements in a given jurisdictions, many are different to accommodate the different needs and expectations of individual projects.

In spite of rising community's demands for greater transparency, under many jurisdictions the terms of special agreements are kept confidential. Where they are not or where the terms are leaked, they create an expectation in the mind of new project proponents that they constitute a precedent and that they should therefore also apply to them on the ground of fairness, which of course defies the very rationale for having 'special' agreements to cater for the special needs of individual projects in the first instance.

The web site https://resourcecontracts.org/ provides access to the "Resource Contracts: A Directory of Petroleum and Mineral Contracts" database, which currently (2019) contains some 2188 contracts negotiated in 92 different countries, with 162 documents detailing relevant tax incentives and exemptions. This web site is developed is the outcome of a partnership of the World Bank, with the Natural Resource Governance Institute (NRGI), and the Columbia Centre on Sustainable Investment, with the contributions of a wide array of other organizations.

While introduced primarily at the behest of mines developers, special agreements also provide government with an opportunity to better spell out some of the obligations to be imposed on the mining company. These may specify any obligation to progress to a given level of downstream processing over a given time scale, to make community development and social infrastructure contributions, to establish appropriate marketing procedures and to provide adequate and timely related reporting and information as well as dispute resolution mechanisms.

Generalising, an increase in tax revenue payable to the State is often impeded by existing stability agreements, which essentially "lock-in" a specific fiscal system. As a result, both good and the bad characteristics apply, and will not adjust with changes in external factors. By way of example, if a tax holiday is guaranteed under a stability agreement, then no tax will be paid during the tax holiday period, despite increases in mineral prices or a weakening currency.

Due to the nature of mining, with long-life projects and large up front investments, providing stability is an important component for attracting investment. The administrating authority needs to understand the nature of the agreement, and model the potential consequences under various potential changes to external factors, including reopener triggers that can help keep the stability agreement in line with its original intent.

From a mining investor's point of view, the most important conditions designed to provide investment stability and predictability include:

- clear and secure land tenure arrangements, with no risk of expropriation or nationalisation of projects and assets;
- fiscal stability in terms of royalties, corporate income tax and other imposts, including clarity around various import and export duties, repatriation of profits etc.;
- immunity from any changes to laws regarding requirements for government and/or local participation in projects and any change regarding specific, temporary or permanent moratoriums, investment incentives and other allowances already granted;
- no export restrictions for processed or partially beneficiated mineral products or commodities, or at least no unexpected changes to existing restrictions; and
- beneficial modifications in the application of other laws which may be deemed necessary to facilitate development, e.g. laws regulating the utilisation of expatriate labour and local labour, including securing working visas for the former group, rights to build infrastructure (port, rail), etc.

From the government's point of view, special agreements provide the opportunity to formalise and better specify their requirements in a number of areas including:

- clarity about any industry commitment to contribute to common-use infrastructure and other necessary community development needs;
- requirements for specified levels of local employment and local procurement;
- obligations to proceed with increased level of downstream processing over time;
- commitment to achieve the best possible price for the mineral product sold and, if sold to a related party, for their transfer price to be demonstrably at-arm-length. This may include requirements for specific marketing documentation relating to cross-border mineral sales; and
- approval procedures for possible future change of project scope, such as a mine expansion and/or partial or total disposal of equity in the project.

While the above lists are not all-inclusive, they provide the main factors, other than the geological and technical merit of a project that may influence an investment decision. These decision-factors will be considered in addition to the standard investment decision factors that include profitability, returns on investment, commodity markets and other typical economic projections and forecasts.

Once set, the terms and conditions of special agreements can only be changed by mutual agreement of the parties. This process may be very time-consuming, complex and cumbersome because renegotiation is generally brought about to satisfy an emerging need on the side of one of the party, either industry or government, which may not represent a priority or even be desirable from the point of view of the other party. Consequently the party initiating the negotiations is generally forced to make significant concessions in return which often makes reaching agreement very hard, if not impossible. In addition, if the agreement has been ratified as an Act of Parliament, any changes agreed by the parties must be submitted to Parliament for approval, followed by complex legislative amendments.

Reality is that neither government nor industry has the capacity to think realistically over the long time horizons of special agreements and that many unforeseen economic and social issues necessitating changes to an agreement are likely to arise under the highly uncertain circumstances which characterise the mining industry, which, if unresolved, have the potential to jeopardise the success of a project from both the industry's and government's point of view.

To circumvent this type of difficulty, modern special agreements generally include the provisions for their renegotiation at specific intervals of time, say every 10–15 years. In addition they often include specific 'renegotiation triggers' in the event the consequences of unforeseen economic, operational or other circumstances exceeding certain predetermined levels, as for instance renegotiation of the distribution of rents if commodity prices leverage company profits above agreed thresholds.

The inability to unilaterally change laws impacting regulatory stability, may be particularly frustrating to an uninformed new government gaining power after democratic elections or a coup even though changes in economic conditions and/or other material events would otherwise justify amending the agreement.

The renewal anniversary of exploitation permits may, unless specifically covered in agreements, introduce the risk that the government may take the opportunity to extract further concession from the company, under an implicit threat that if these are not satisfied, renewal of the licenses may not go ahead. There are many circumstances where renewal of a mining title may become necessary as for instance conversion of an exploration permit into a mining or exploitation permit after completion of various economic, technical, environmental, social and other studies.

8.2 Security of Tenure

The most important aspect of any resources investment potential is the right to hold and retain, without the risk of losing, the access and associated legal right to exploit the minerals being the subject of the investment.

Prior to any initial investment being made into a country, a potential investors will be wise to seriously consider the sovereign risk of that investment destination.

At a macro level sovereign risk is generally a reflection of the risk that a country's central bank through the government may default on its due repayments of international (sovereign) debt or of not honouring a loan agreement. In addition to debt repayment, sovereign risk also reflects the risk that the government makes decisions or takes actions that negatively impact on the economic welfare of that nation. These generally manifest as changes to rules and regulations that may negatively impact on investment outcomes unless existing investments can be quarantined from them ideally through stabilisation agreements or grandfathering.

The timing of negotiating around retention of tenure to minimise the risk of loss is upfront, that is to say as soon as possible after the decision to invest is made. Engaging early with the relevant government authorities is important because once the company has been granted an exploration permit, it is obliged to spend a specified, minimum amount on exploration activities each year as well as incurring the payment of tenement fees both fixed and/or dependent on the area held under the license, often referred to as ‘dead rent’. Significant amounts of money will be flowing out of the company to the government and to the consultants doing the exploration prior to any potential mineral discovery and related production income.

In this light, it would be very wise for the mining company to negotiate, at least in principle, an agreement with government around investment in further exploration and later in mine development. This should include a preferential right for the discoverer of a new mineral deposit to be automatically granted a mining license for its exploitation, unless such right is already embodied in the applicable Mining Act.

Consider the case in which no such discussions have been held and the company has spent $20 million over 5 years on broad-acreage exploring in the country. Given the general requirement that government the company should provide at least one detailed report on its exploration activities at a minimum annually, it means that the government has detailed information about all or most of their detailed exploration results, which carry significant value.

Once the exploration programme has been completed and assuming that the various studies have located a deposit likely to be economically attractive, the company will have to apply for an exploitation or mining permit, If the company had not previously engaged with the government to establish the investment rules to transition from exploration to development, then at this point the government has an opportunity to impose, without negotiation, a range of rules and regulations designed to ensure that it receives as high a share of profit as the project may bear in the knowledge the company is now captive to its project. Under this set of circumstance government may exploit the implicit threat of denying the award of the exploitation/mining permit unless the company agrees to all its requests. Stated differently, the company borne the full risk of exploration but lost its negotiating power and is at risk of losing the concession if it does not agree to what may amount to unfair impositions. Of significant importance is also the right for an exploration company to be able to dispose of its exploration rights in part, as for instance by means of a joint venture or in full by means of sale. Past history clearly indicates that many mineral discoveries are made by small to medium- size explorers (SMEs) which invariably lack the significant financial resources and expertise necessary to develop them. Many SMEs, exploration programs are financed by major mining companies under a commitment that, should a major mineral deposit be discovered, the major has a right to acquire a predetermined level of equity in the project for an agreed price, which may be determined by a number of different mechanisms. This type of deal is attractive to the majors because on average SMEs are more efficient and effective explorers due to their flatter and more agile management and much lower level of overheads.

8.3 Transparent and Fixed Fiscal Goal Posts

Prior to an investment decision being taken, the investor will assess the political and economic stability of the country in which the investment is to be made. The investment ranking or risk of a country can typically be sourced from any of the three major credit rating agencies being:

- Standard & Poor's (S&P) (United States);
- Moody's (United States); and
- Fitch Group (New York and London).

There are additional rating agencies, including Morningstar Credit Ratings which is gradually increasing its market share.

An idea of a country's risk can also be derived from the risk-categorisation table that is regularly and freely provided by the OECD (2019) for use in setting risk premia, for instance for cross-boundary settlements. The OECD country risk table, reproduced in Appendix B, includes eight categories of which Category 0 covers the most stable, primarily developed economies, while Category 7 covers those with the highest country risk, many of which are often the theatre of conflict. The political and economic stability of a country is a necessary assessment because the investment capital actually belongs to shareholders and other stakeholders and not to the company's management team. The latter is mandated to ensure that the investment meets the requirements of all stakeholders and that the secured capital is invested responsibly and with due regard to their appetite for risk.

The fact that the investment capital actually belongs to investors and not to the company itself is often a critically overlooked factor. This is because when the government imposes new or varies the economic rules applying to existing or future operations, investors have the option to redirect their investment capital to an alternative company if they deem the jurisdictional risk is too high in terms of their risk profile.

Government must also be acutely aware of sovereign risk driving away investment capital if their policies are not transparent, out of line with other competitive countries and above all if they vary the fiscal goal posts *ad hoc*.

It is very difficult to convince investors to once again invest in a specific jurisdiction that has a reputation of having changed the fiscal goal posts on previous occasions. Rebuilding investors' confidence is indeed a significantly harder task than destroying it.

Many jurisdictions that may be deemed to be high-risk investment destinations may be afforded Political Risk Insurance (PRI) albeit at a price. PRI may be provided by a number of insurance organisations such as Lloyds of London, AIG, Aon and others, and cover various politically-induced risks including:

- war or civil unrest;
- cancellation or withdrawal of permits or licenses;
- export or import limitations and restrictions;
- termination of Government purchaser contracts;
- transfer, foreign exchange and payment restrictions;
- Government intervention;
- reneging of Government contracts;
- nationalisation or confiscation by a Government or associated body;
- expropriation of equipment; and
- banning rights of disposal of assets.

8.4 Purposes and impacts of Stabilisation Agreements

Intuitively, stabilisation agreements exist to provide potential investors with certainty or to mitigate uncertainty relating to specific risk factors that may otherwise negatively impact their investment decision. These agreements are entered into to provide both the Government and the investor rules around investing in his/her specific project in the country and a commitment that they will not be varied over a fixed period.

Investors view stabilisation agreements in a positive light since they capture and disclose Government guarantees around not changing certain regulations and investment conditions. The implications of being able to secure a stabilisation agreement with a specific Government is positive and will tend to attract investors into the country which may otherwise have been initially perceived as too risky to invest in in the absence of such an agreement in place.

In general, a stabilisation agreement will not be available from the Government if the expected investment quantum is relatively small and/or the expected life of project is short. However, if the capital required to develop the project is significant or if the project itself is deemed to be a strategic project by the Government, the agreement may be made available.

8.5 Risk of Stabilisation Agreements Being Tampered With

Although as already discussed no element of a stabilisation agreement can be altered unilaterally by government without the agreement of the industry party, there have been situations in some developing countries, particularly in Africa, for the Government to renege on established agreement in part or as a whole.

In the event that a stabilisation agreement is tampered with in any way, a significant impact will be the notable and immediate reduction in new foreign direct investment into the country or into the particular sector to which the agreement applied. The government will be held directly responsible for creating this investment uncertainty and instability and the repercussions will be economically and politically significant.

In addition, for those investors already invested in that country, a review of their investment exposure and a reassessment of their potential investment risk will likely be conducted. The decision to remain invested or to further invest through possibly recapitalising existing projects, production expansions, upgrades or additional exploration, may be compromised and ultimately discontinued if their own or another investor's stability agreement is reneged upon or unilaterally adjusted or tampered with. It is also possible that an existing operator may place its operations on care and maintenance subject to further assurances that their operating conditions do not adversely change.

Very often the trigger precipitating this type of crisis is the political pressure by a growing popular perception during periods of mineral commodity boom that government revenue is not leverage to the same extent than company profits by increasing commodity prices. Mounting government pressure for the renegotiation of special agreements is generally resisted by industry which is under no obligation to forego part its windfall profit that it views as compensation for long previous period of operation with very lacklustre results.

An interesting case study is that of a prominent African copper producing country that found itself in these frustrating circumstances, claiming, with a degree of justification, that the original stability agreement entered into with some of its major copper producers were negotiated under duress while the industry was on its knees and with an inadequate understanding of its long-term consequences. It also realised that mining royalties were harder to avoid than corporate income tax, having engaged in long-protracted, inconclusive and acrimonious litigation involving allegations of industry abuse of transfer pricing and misinvoicing.

Under mounting political pressure brought about by a popular perception that industry was getting an unfair share of rapidly climbing rents during a recent commodity price boom, the government, unilaterally and with essentially no consultation, declared the special agreements null and void, abolished corporate income tax for mining and increased the royalty rate many fold, particularly for open cut mines.

The industry response was immediate and fierce. Significant, already-approved investment in the expansion of current operations and in increasing downstream processing were halted resulting in the loss of thousands of direct mining and multiplier jobs throughout the economy. A review of the future profitability of some of the labour-intensive underground operations also led to some closures and significant cost cutting measure resulting in further job losses. In addition, the mining industry embarked on a significant public relations campaign highlighting, on the eve of a national election, how this hardship had been uncalled for and totally attributable to poorly conceived government initiatives. The industry also considered the unprecedented possibility of taking government to an international court of arbitration. The consequences of potentially many years of legal uncertainty and particularly of the likelihood of a judgement against the government arose the spectre of the latter not being in a position to pay the likely level of compensation and becoming insolvent.

Over the period, the government that introduced these draconian rules lost the next elections. The new government moved rapidly to revoke all the previous legal amendments and re-established the *status-quo*, albeit securing some minor concessions in terms of royalty rates. Industry proved fairly timid in resuming their expansive initiatives, which was not helped by commodity prices becoming softer the meanwhile.

In the end, the whole episode was highly disruptive and value destroying for both parties and exacerbated the level of distrust and general tendency for government and industry not to consult and communicate effectively and to deal in most instances in an adversarial manner.

In another African country, the government levied an accusation at an international mining company operating within its borders accusing it of transfer pricing and that the operation had consequently failed to pay many billions of dollars in due taxes over time. This was naturally disputed by the mining company, but the issue also raises areas of concern. Transfer pricing is a system whereby products are "sold" to an offshore entity typically owned by the same owner of the primary producer of that product, at a price that is less than a fair market price. For example, a mining company may sell its unbeneficiated ore sourced in one country to its own beneficiating facility in another country, where each country has a different tax structure, or possibly where infrastructure in one country is better (cheaper to use) than in the other. This means that the sales revenue received by the producer is lower than it should be and that, after deducting operating costs, there is little taxable profit remaining. On the other hand, the country receiving and thereafter further processing the ore into a higher value product will recognise a disproportionately higher sales revenue from the product (once on-sold to a consumer). Consequently, tax realised by this latter State will be higher than what would otherwise be construed as fair.

Some of the above actions may also result in the event that a government changes the investment environment in the absence of a stability agreement. An example of this occurred when a different Sub-Saharan African country decided to promulgate a new super profit tax over its copper producers. This change resulted in the actual closure and further threats to close existing mining and processing operations in that country. The super profit tax was rescinded, but significant economic damage had already been done and the country was immediately deemed to be a less favourable investment destination due to increased sovereign risk.

Take away points

Government's point of view

- The use of Special or Stability Agreements should be limited, if possible, to economically or strategically important projects necessitating the inclusion of a number of out of the norm conditions and/or obligations on the side of both government and industry. All other projects should progress under the terms of the prevailing mining and tax legislation.

- Agreements should not be indefinite over the life of the project but should cover defined shorter (e.g. 10–15 years) periods if necessary renewable subject to negotiation. They should also include 'renegotiation triggers' when certain parameters (e.g. commodity prices) fall outside a predetermined range.
- Agreements, particularly if ratified by Parliament, should be public documents. However this creates 'precedent' and an expectation on the side of new mine developers that any existing reduced royalty or tax rate should be the ceiling for future negotiations.
- Agreements should include an obligation on the side of the company to promptly submit on request documentation such as invoices for sales conducted abroad and for marketing, shipping and related costs incurred beyond the national borders.

Mining industry's point of view

- Given the degree of a country's risk a Stability Agreement may be a prerequisite condition to investment in a mining project.
- Depending on the nature of the obligations entered into on the side of the company, the longest term, ideally over the life of the project, would be desirable.
- Ratification by Parliament would be more secure even though any future amendment would involve more complex procedures than a simple renegotiation with a relevant Minister.
- The degree to which a stability agreement could be contested by industry in an international court is unclear even though there have been numerous serious breaches on the side of some developing countries' governments.
- The trend for shorter terms and the inclusion of renegotiation triggers is rapidly being established and industry will inevitably have to learn how to conduct itself in this new environment.

References

Mining Tax Incentives Database (2019). https://www.bing.com/search?q=mining+tax+incentives+database+2019&form=EDNTHT&mkt=en-u&httpsmsn=1&msnews=1&plvar=0&refig=123ca62dfade4766b6dbfdceeb33e9e1&sp=-1&ghc=1&pq=mining+tax+ince&sc=0-15&qs=n&sk=&cvid=123ca62dfade4766b6dbfdceeb33e9e1

OECD (2019) Country risk classifications of the participants to the arrangement on officially supported export credits. http://www.oecd.org/trade/topics/export-credits/documents/cre-crc-current-english.pdf

Resource Contracts: A Directory of Petroleum and Mineral Contracts (2019). https://resourcecontracts.org/

Chapter 9
Administering and Complying Fiscal Regimes in a Globalised Mining Industry

Abstract There is a vast difference between formulating sound fiscal mineral policy/legislation and successfully administering it in a manner that promotes the desired taxpayer behaviour, facilitates compliance and reduces tax minimisation, avoidance and the incidence of disputes. The progressive globalisation of the mining industry has created a new modus operandi including fragmentation of the value chain with increased cross-border transactions between related companies owned by the same multinational enterprise (MNE). Significant issues arise to ensure that the related transfer prices are at-arms-length and ideally result in the appropriate level of profit and tax being paid in the jurisdictions in which value has actually been added with no erosion of the tax base of and shifting of profits from the country hosting the mining operations. In spite of significant efforts by international institutions, such as the OECD, the UN, The IMF and the World Bank, developing appropriate guidelines and governments amending their tax systems to close emerging loop holes, MNEs continuously restructure their operations to take advantage of tax arbitrage opportunities to pay most of their taxes in foreign, often low tax jurisdictions thus minimising the amount of tax paid at the consolidated enterprise level. These processes impose considerable administrative pressure on both generally under resourced tax authorities and industry.

Keywords Administration · Multinational · Taxation · Returns

9.1 Fiscal Policy, Legislative and Administrative Frameworks

For a government mining fiscal policy to be successful in influencing taxpayers' behaviour in a manner that achieves its objectives, as discussed in Chap. 3, it must be supported by an effective legislative and regulatory framework and, above all, by adequate administrative capacity to enforce it.

It is critical that a country's constitution and/or its primary fiscal legislation should have unambiguous powers to establish subsidiary legislation and

E. Lilford, P. Guj, *Mining Taxation*, Modern Approaches in Solid Earth Sciences 18,
https://doi.org/10.1007/978-3-030-49821-4_9

administrative rules and regulations and to confer, beyond legal dispute, to both central and decentralised levels of government their legitimacy to levy various types of taxes.

To the extent that the taxation legislation imposing corporate income tax across all sectors of the economy, including mining, must operate alongside other pieces of legislation designed to levy other forms of mining taxes, such as mineral royalties most frequently levied under a mining act and related regulations, it is crucial that it should dove tail well with them. More often than not, to facilitate this, the head taxation legislation includes a section specifically devoted to the mining industry.

Ideally the fiscal legislation and supporting regulations applying to mining should be as clear and unambiguous as possible to limit opportunities for unwarranted tax minimisation and avoidance, and to reduce the incidence of disputes without imposing an excessive compliance burden on both industry and government.

The procedures for assessing and impose payment of various mining taxes should also be well defined to reduce the degree of ministerial or agency discretion, often the source of corrupt practices, to the minimum administrative level necessary to ensure practical and smooth enforcement.

Legislation should also include dispute resolution procedures and mechanisms and penalties commensurate to the infringements, firm but equitable to both government and industry and designed to make use whenever possible of mediation and arbitration, thus averting as far as possible court proceedings as a measure of last resort.

In practice different jurisdictions adopt vastly different mining taxation administrative frameworks that range from highly centralised, particularly in the context of corporate income tax, VAT, import/export taxes and excises, to highly decentralise on a regional basis or on the basis of largely autonomous states/provinces.

In the case of federations (e.g. Australia, Canada, USA, etc.) the power to legislate about and administer matter pertaining to land (including minerals) rest in the majority of cases with individual states/provinces. As a consequence, mineral royalties (and in a minority of cases mining income tax) are generally collected and retained at the individual state/provincial level.

Furthermore, while legislation relating to corporate income tax, VAT, import/export taxes and excises is in the vast majority of cases formulated and administered centrally by the country's federal Ministry for Finance and one or more of its agencies, typically its Tax Office, Custom and Excises, etc., mineral royalties may in many cases be legislated and administered under the prevailing Mining Act and related Regulations by the relevant Department of Mines, either centrally or at the state/provincial level. Effective and efficient enforcement under these different regimes may require very different government and public service structures and protocols to define their respective responsibilities and accountabilities and their interaction often lacks in cooperation and gives rise to communication and coordination issues between the different ministries and agencies involved, in many cases resulting in inefficiency and duplication of effort.

Even in countries with centralised mineral royalties and corporate income tax regimes their administration and tax collection tends to be carried out using

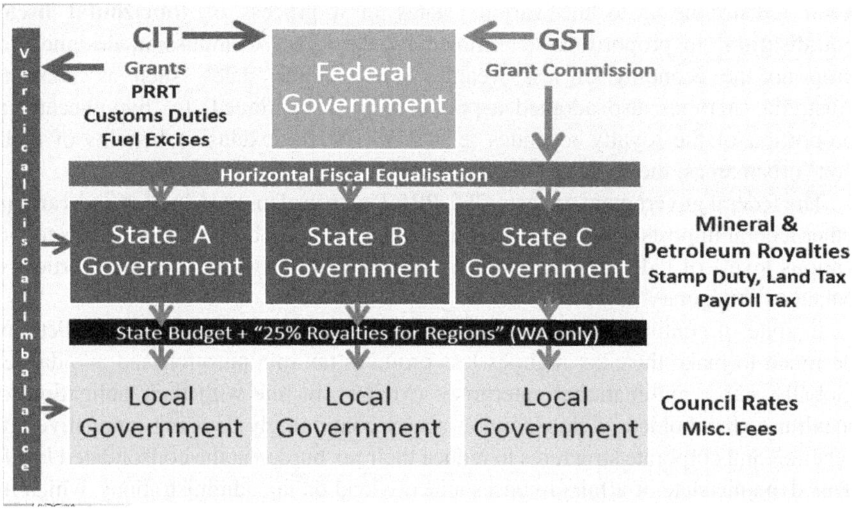

Fig. 9.1 Schematic representation of the fiscal processes for the collection and redistribution of revenues from the main sources of taxes in a federation such as Australia

regionalised structures. In the majority of cases the related revenue is remitted to a central Treasury and appropriated in its Consolidated Revenue for reallocation to various regions according to national government budget priorities. In other cases political pressure dictates that a proportion of the revenue should be 'hypothecated' to the region and/or some of the communities directly affected by the mining operations.

By contrast, in federations revenue from mineral royalties is generally appropriated to the state's/province's Consolidated Revenue and in the majority of cases reallocated to various regions through priorities set in line with the state/provincial budgetary processes. Communities directly affected by mining operations and indigenous traditional landowners are often compensated directly by mining companies for the impact imposed on them, although the link between the degree of disturbance and the amount of compensation often becomes tenuous and compensation may take the form of a royalty on production even though the communities may not be the legal owners of the resources in the ground which generally belong to the state.

To avoid a situation where mineral rich states/provinces in a federation may become disproportionately wealthy by comparison to other states/provinces, policies and processes are generally applied at the central federal government level to redirect funds to ensure that minimum standards of living are maintained in those states/provinces which are not mineral or otherwise endowed and the economy of which would otherwise lag behind.

Figure 9.1 portrays the example of Australia where each state collects and retains, asides from various other state imposts, mineral and petroleum royalties levied on production from their land mass and adjacent territorial waters. The states also collect good and service tax (GST) but remit it to the federal government which good redistribute it to the various states in a process of **'horizontal fiscal equalisation'** in proportions determined by the Grant Commission designed to prop up the economy of less wealthy states. Some states, such as Western Australia, at times also adopted a policy (now discontinued) to 'hypothecate' a proportion of the royalty revenues (e.g. 25%) for the exclusive benefits of their non- urban areas, the so called 'regions'.

The federal government collects CIT, PRRT on petroleum production and various other federal imposts and redistribute the related revenues by means of 'grants to the various levels of federal, state and local government through hopefully 'vertically balanced' budgetary processes.

In spite of continuous process improvements in most mineral-rich jurisdictions designed to make their tax regimes less prone to tax minimisation and avoidance, globally active multinational enterprises (MNEs), in line with their obligation to maximise shareholders' value, counter-act by continuingly devising innovative tax schemes and corporate structures to reduce their tax burden at the consolidated level. This dynamic state of affairs imposes a heavy load on tax administrations, which in many developing countries are inadequately skilled and more often than not under resourced. Indeed tax administration has been identified as one of the most critical areas in need of capacity building and is the focus of significant support from international institutions such as the World Bank (WB), the United Nations (UN), the International Monetary Fund (IMF), the Organisation for Economic Cooperation and Development (OECD), as well as from the foreign aid agencies of many developed countries.

9.2 Mining as a Global Business

9.2.1 Increasing Globalisation of the Mining Industry

The progressive removal of many barriers between different national economies has encouraged a freer flow of trade, investment, capital, goods, services, technology, communication and, to a lesser degree, labour across national borders. The term **'globalisation'** has been applied to this progressive opening of local and nationalistic perspectives to a broader outlook in an interconnected and interdependent world resulting in progressive efficiency-driven industrial restructuring and integration of many aspects of diverse national economies.

Economic reasoning justifies globalisation in that it generates '**comparative advantage**' by promoting each country to specialize in producing and exporting primarily those goods and services which it can produce more efficiently and to rely on importing goods and services which it cannot produce competitively. Thus, in theory at least, free trade should be beneficial to all countries as long as they exploit their individual comparative advantages as this should lead to more efficient markets, increased competition and productivity and more equitable spreading of wealth.

A country's comparative advantage depends on its endowments in terms of its factors of production (land, which includes mineral resources, capital, and labour). Many developing countries are rich in minerals and unskilled labour, but lack in technology, entrepreneurship and above all in domestic capital adequate to fuel their development. In this context globalisation has fostered a rapid increase in foreign direct investment and technology transfer in the area of mineral exploration and mining in developing countries, while at the same time enabling large MNEs to achieve economies of scale, reducing their cost of production and to grow.

On the other side of the coin, some contend that globalisation by fostering the rise of powerful multinational or global firms and other international organisations may undermine national sovereignty and independence. Furthermore the benefits of globalization may be unfairly skewed towards rich nations or individuals, creating a sense of inequity and potentially giving rise to nationalistic or even xenophobic sentiments and possible conflict.

Globalization has in many instances also led to economic decisions involving company strategy such as expanding, downsizing or closing down being shifted away from local control and being made with inadequate reliance on the views of local managers and consideration of local impact. Rodrik (2017) contends that there may be a need to 'rebalance globalisation' to take into greater consideration the needs and expectations of labour for jobs and income stability. Recent tendencies towards re-emergence of protectionism are a political expression of this discontent.

While the upstream components of the mining value chain (e.g. mining and initial mineral concentration) are retained at the mine site in the developing countries where the mineral deposits occur, much of the high value-adding downstream processing, marketing, financing, corporate and technical services tend to be provided primarily by associated companies located abroad, which are part of the same MNE group. The result is that **an increasing proportion of mining transactions are carried out cross-border between related entities**. This fragmentation of the mining value chain has, as discussed below, very significant tax implication and some economists suggest that cross-border investments in recent years have not necessarily been made to build capital infrastructure and operational and marketing hubs in the most logistically suitable countries but to seek, as schematised in Fig. 9.2, countries with the lowest taxes and/or favourable double taxation agreements/treaties.

9.2.2 Taxation Issues Created by Globalisation of The Mining Industry

Mining MNEs have tended to structure their businesses by consolidating high-value functions and related intangible assets, in **service centres or hubs** which provide goods and services to their regional or global operations.

While in many instances this is a legitimate endeavour to maximise shareholders' value by achieving critical mass through pooling of specialised resources, proximity

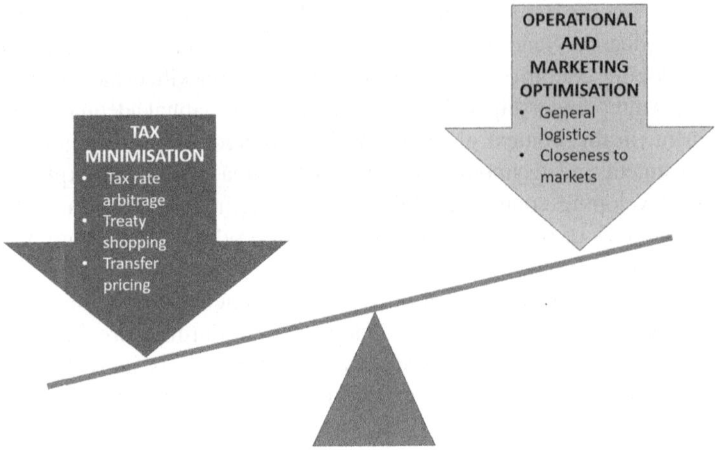

Fig. 9.2 Relative importance of fiscal and operational considerations in MNEs' corporate restructuring decisions

to customers, trading and shipping centres and/or research facilities, as well to attracting and retaining highly skilled personnel by stationing them in amenable locations, an undesirable consequence is that the functions of MNEs' mining subsidiaries are often stripped down to mostly routine activities utilising primarily less skilled personnel and tangible assets.

There are, however, many instances where service hubs are patently being located in low-tax jurisdictions to take advantage of:

- **inherently lower corporate tax rates**; or of
- the country allowing the establishment of **preferentially-taxed special purpose entities (SPEs)**; or of
- **reduced or no withholding taxes applying by virtue of tax treaties such as a Double Taxation Agreement (DTA)** with the nation hosting the mining operations.

To ensure that tax is paid only in jurisdictions where value has actually been added, in effect to **prevent eroding the tax base of the country hosting the mining operations by shifting profits abroad**, the tax authority must ascertain whether:

- the profits of mining subsidiaries and of overseas related customers and/or service providers match the value actually added by each of them along the mining value chain, that is to say whether the economic substance and the form of an arrangement do match; and
- the price adopted for these goods and services provided by associated companies, the so-called **'transfer price', was set at 'arms-length'**. That is to say as if each transaction had taken place between unrelated parties in a contestable market, with the price fairly reflecting the nature of the **function(s) performed, the assets utilised and the risk taken by the provider of the good and services**.

The complexity of correctly setting arms-length transfer prices in accordance with the five methods provided in the Organization for Economic Cooperation and Development (OECD) *Transfer Pricing Guidelines for Multinational Enterprises and Tax Administrations,* 2012, as summarized in Appendix E, has created unbearable administrative pressure on industry and the tax authorities in the countries hosting the mining operations and is an avenue for both honest **mispricing** and opportunities for some less scrupulous taxpayers to abuse the system. Mispricing, which may arise from:

- **Undercharging for mineral products exported** and transferred to related parties. This risk frequently arises because few mining companies are fully vertically integrated and mostly export to related, smelters or marketing hubs crushed and screened ore (as for instance iron ore, bauxite and coal), or base metals and other concentrates or intermediate products after limited processing; and
- **Overpayment for goods and services** increasingly provided by related parties including a range of both routine (e.g. most corporate services) and specialised (e.g. marketing, treasury/financing, insurance, logistics, engineering, metallurgical and technical/R & D) goods and services.

Logically, mining companies, acting independently, would be reluctant to accept excessive discounts on their mineral products sales and/or pay grossly above market prices for marketing, finance (including interest and fees on borrowing and guarantees}, technical/R&D, corporate and other goods and services, if they were obtainable at a lower price from an unrelated supplier. They would also be unlikely to divest of assets, at times below their market values, to a related entity and subsequently pay that entity premium fees and/or royalties for their use.

Mispricing reduces the profit of the mining subsidiary and therefore the tax collected in the host country. From a tax authority point of view, particularly an under-resourced and under-skilled one in a developing country, auditing of transfer prices is a very complex process hindered by the paucity and general irrelevance of available **'comparables'** and by the general reluctance of MNEs to disclose information about relevant expenditures incurred in foreign jurisdictions beyond the reach of the local tax authority. This situation, however, is improving as an increasing number of DTAs include provisions for exchange of tax information between the signatory countries.

Mispricing differs from the criminal practice of **misinvoicing**, which involves, for instance, understating the quantity and above all the grade and therefore the value of exported mineral products. There have been recently alarming estimates of the revenue lost for instance to African countries due to illegal trading including misinvoicing including serious allegations about specific mining companies engaging in misinvoicing with regard to their exports of mineral concentrates in some African jurisdictions. These allegations have invariably been strenuously denied by the companies in question the practices of which are currently under investigation.

These practices, while hopefully confined to a minority of companies, do nonetheless work very much to the detriment of law abiding member of the mining industry and the general tenor of their rapport with local taxation authorities. Mispricing and misinvoicing have also justified in the eyes of politician and tax authorities the setting of sometimes draconian pre-empting **anti-avoidance measures** and **high withholding taxes as a first line of defence**.

9.2.3 Reform and Compliance Challenges in a Dynamic Taxation Environment

Following the 2012 G20 meeting the OECD was tasked with formulating a series of policies and strategies to counteract the tax-effective schemes and structures adopted by MNEs to exploit gaps and mismatches between the tax rules of different countries. To the extent that the industry's posturing has the effect of eroding the tax base of and artificially shifting profits from some of the countries hosting mining projects to low or no-tax jurisdictions where there is little or no real economic activity, the OECD aptly named this initiative BEPS an acronym standing for Base Erosion and Profit Shifting. After wide and comprehensive consultation, a **BEPS Action Plan**, including 15 specific initiatives, was formulated and released in late 2015 (OECD 2015). The title and nature of each specific action are listed in Appendix F.

Adherence with some of the BEPS Actions will require many of the 44 OECD countries and numerous other voluntary participants to draft and enforce appropriate domestic legislation. This transition will take time, but ultimately it is expected that all the recommended Actions will be broadly acted upon and have a very significant influence on most globalised mining operations. These changes will without doubt create major challenges and compliance costs for multinational mining companies and over time increase their tax burden and decrease their ROI at the consolidated level.

In February 2019 the OECD also released a consultation document, the so-called BEPS 2.0, to address the challenges of taxation in the digitalizing economy. This was followed by public consultation held in Paris the following March. The proposed policy include two pillars. Pillar 1 seeks to determine whether more profit should be recognised and taxed in countries making significant use of digital services and/or intangibles. Pillar 2, which follows the approach of the U.S. Global Intangible Low Tax Income (GILTI) policy, is a global anti-base erosion proposal that would set a minimum effective tax rate in response to concerns that profits from intangible assets are often subject to no or very low rates of taxation. While it will take a number of years to reach consensus and better define the related tax reform, it would appear likely that a possible outcome will be that MNEs, which are producers and exporters of mineral commodities, may have to pay a higher level of tax at the consolidated level.

9.2.4 *Tax Minimising MNEs Structures and 'Treaty Shopping'*

BEPS Action 6 is designed to **counter tax treaty abuse**, where it can be demonstrated that MNEs have arranged their affairs and corporate structures for the purpose of deriving benefits under a tax treaty, such as a double taxation agreement (DTA), in a manner which was not intended by their signatories. To prevent double taxation, DTAs often provide for significant reductions or even elimination of payment of withholding tax on the remittance of dividends, interest and other payments otherwise imposed by the country in which the mining company is resident. It is, therefore, not surprising that MNEs generally do not invest directly in mining projects in typically high-tax countries, but hold them through subsidiary 'conduit' companies registered in low-tax jurisdictions which have a DTA with the country hosting the mining operations, as shown in its simplest structural form in Fig. 9.3.

As already discussed, mining companies generally provide a range of non-tax-related justifications for their corporate structuring. While these may in some case have some validity, it would appear difficult, for example, to explain how a MNE holding a project in Namibia through a conduit company in the Mauritius Island, justifies holding a similar project in Malawi through a subsidiary in the Netherlands other than for the taxation advantages of these countries having DTA with them and with the country of residence of the MNE.

BEPS Action 6 recommendations is that tax treaties such as DTAs should contain one or a combination of "minimum standard" anti-abuse provisions such as:

- a **limitation of benefits (LOB) test** requiring the entity to have a substantial presence in the territory of residence; or
- a **principal purpose test (PPT)** whereby no DTA benefit can be derived if it can be proved that the motivation for establishing a business in the jurisdiction was primarily to derive the relevant DTA benefits.

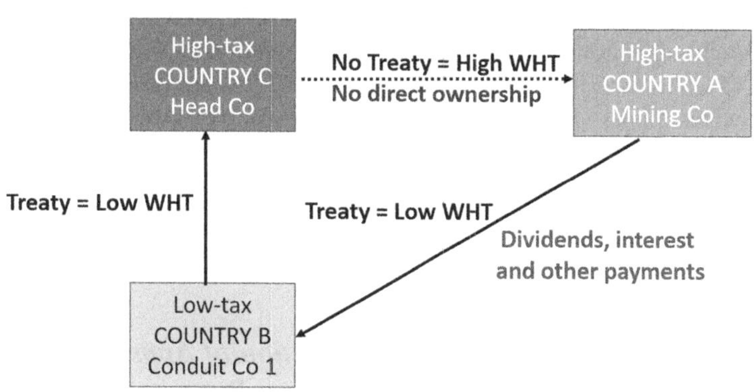

Fig. 9.3 Simple corporate structure designed to minimise withholding tax payments

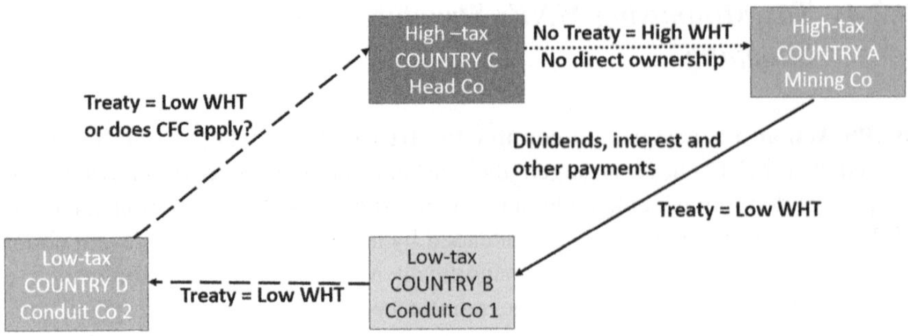

Fig. 9.4 Examples of multi-layered treaty shopping (Modified from IMF 2014)

The extreme complexity and artificiality of some of the multi-layered structures adopted to achieve **'tax rate arbitrage'**, a practice referred to as **'treaty shopping'** (IMF 2014), together with the fact that many of the related 'conduit' companies are effectively just 'mail boxes' with no appreciable staff or clear business purpose, adding little or no value, would indicate that these subsidiaries are, in most cases, primarily designed to minimise MNEs' tax liabilities at the consolidated level.

For example, as illustrated in Fig. 9.4 (modified from IMF 2014), in the absence of a treaty between country B and C, the Head Co. may make use of a second Conduit Co. 2 resident in Country D, which has tax treaties with both Country B and C. Proposed changes in the rules relating to controlled foreign companies (CFC), however, may also hinder some of these arrangements in the future.

The speed with which some MNEs change their structures to counteract the closure of fiscal loopholes by governments, also corroborate the view that tax considerations are the predominant criterion for their international corporate structuring. While these tax practices may be technically legal, it may be argued they are ethically questionable.

It is expected that Action 6 will have significant implications and will precipitate major restructuring for the vast majority of mining MNEs. It will also be a mammoth effort for countries to amend their DTAs accordingly. To facilitate the task **BEPS Action 15** led to the establishment from first July 2018 of a **multilateral instrument** (MIS) which allows governments to modify existing bilateral tax treaties in a synchronised and efficient manner to quickly implement treaty-related BEPS recommendations without the need to expend resources renegotiating each treaty bilaterally. To date over 100 jurisdictions have agreed to participate in this initiative.

9.2.5 Issues Relating to the Determination of Appropriate Levels of Debt and Related Interest Rates

Of particular interest is **BEPS Action 4** which seeks to prevent MNEs from concentrating their **interest deductions** in the typically high tax jurisdictions hosting

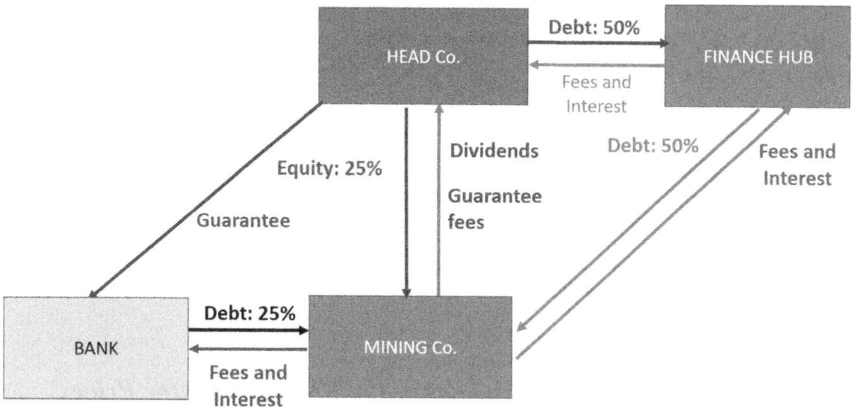

Fig. 9.5 Typical Treasury/Financial Hub structure providing funds to a mining company in a jurisdiction with 3:1 thin capitalization rules (modified after Guj et al. 2017)

their mining projects by extending to their mining subsidiaries through their financial hubs:

- levels of debt finance well in excess of what is customary for the mining industry (i.e. around 40%) sometimes up to 100% of the relevant capital cost of projects or to the maximum allowed if **thin capitalisation** rules apply (typically 66–75%);
- at interest rates significantly in excess of those incurred on borrowings by their financial hubs which are subsequently on-lent to their mining subsidiaries; as well as
- charging significant banking and guarantee fees on the above loans.

In the example of Fig. 9.5 the mining project is funded in accordance with the prevailing 3:1 thin capitalization rule. The Mining Co. remits to the Head Co. dividends on its 25% equity funding with the 75% debt component being provided 50% by the finance hub and 25% by an unrelated bank. The bank loan is secured by a guaranteed by the Head Co. for which it charges the Mining Co. a hefty fee. The remaining 50% of funds are provided in the form of an internal loan from the related Finance Hub. Head Co, which borrows at low rate of interest and on-lends to the Finance Hub in return for an interest margin and loan fees. The Finance Hub in turn on-lends to the Mining Co., receives a lending fee and justifies a lofty increase in interest rate on the basis of increased risk exposure. Both the high fees and interest payments to the unrelated bank and to the foreign resident related Finance Hub are claimed as a legitimate deduction against taxable income of Mining Co. and, have the effect of dramatically reducing its profit and its capacity to pay dividends including to government if it holds an equity share in Mining Co.

Re-characterisations of debt to equity and adjustment of applicable interest rates have recently been imposed by the courts to this type of transactions particularly in the case of petroleum developments, resulting in the imposition of significant backdated tax payments and penalties after long protracted and costly legal battles.

The difficulty in achieving early resolution is often attributable to the fact that, particularly in the case of developing country with poorly developed banking systems and low domestic capital generation, determining an appropriate rate of interest adequately compensating for the specific combination of project and country risk often proves to be an insurmountable hurdle.

To avoid this type of issues the Action 4 recommendation is that countries should amend their domestic tax rules to limit interest deductions by a taxable entity to a specified percentage in the range of 10-30 percent of earnings before interest, tax, depreciation and amortization (EBITDA).

9.2.6 Issues Relating to the Determination of Transfer Prices on the Provision of Goods and Services

BEPS Actions 8–10, which seek to align **Transfer Pricing Outcomes with Value Creation**, will affect the determination of transfer prices in a very wide range of controlled transactions between related parties typical of the mining industry, These typically involve the transfer/sale of a variety of mineral products, the provision of technical/engineering and R&D services, involving specialised skills and hard to value licence fees/royalties for the utilisation of proprietary intangible assets (Action 8), and of more routine corporate services including HR, accounting, self-insurance, legal, training, etc.

Action 9 covers compensation relating to the contractual allocation of the risk of various activities to the various related parties and its impact on the allocation of profits to them. It also addresses the appropriateness of returns on equity capital provided MNE group members.

Action 10 focuses on profit allocated to transactions which appear not to be commercially justifiable or rational. In cases where these transactions can be proved not to be aligned with value creation and to divert profits from and erode the tax base from jurisdictions where the most economically important activities of the MNE group were carried out, they will be subject to **re-characterisation** to reflect the conditions that would have been agreed between unrelated parties in uncontrolled transactions. Taxpayers' claims that re-characterisation should be avoided because it may results in double-taxation, may be exaggerated because most jurisdictions provide foreign taxation credits or exemptions. The UK has introduced, and more recently Australia has announced, a 'diverted profits tax' to more effectively address these issues.

Marketing of mineral products is a critical area in the context of transfer pricing, particularly from the point of view of government, given the significance of the revenues involved, the fact that many MNEs have elected to provide marketing services through related **marketing hubs** and the complexity and opacity of the markets and selling processes of many mineral products.

Sales/transfers may involve a range of different mineral products along the downstream processing route. These range from crushed and screened ore (e.g. iron ore, bauxite, coal) through intermediate products, such as mineral concentrates (e.g. precious and base metals concentrates, heavy minerals sands), and/or partially chemically processed products (e.g. alumina, lithium carbonate, vanadium pentoxide, etc.), to various essentially smelted (e.g. gold/silver dore', blister copper, nickel matte, etc.) and/or refined metallic forms (e.g. gold, copper, nickel, etc.). Pricing for smelted and refined precious and base metals sold on the basis of stringent quality standards on well-developed international terminal markets (e.g. the London Bullion Market (LBM) and the London Metals Exchange (LME)) does not generally create difficulties. By contrast pricing for most of the other mineral commodities and products for which no established terminal market exist can create significant and at time insurmountable challenges. In some cases, such as for iron ore, the recent development of reliable and regularly published pricing indices for various iron grades together with discounts for various impurities, such as Platt's IODEX, helped overcome the difficulty. Many other mineral products, however, are sold in poorly documented and opaque markets where differences in quality and/or value-in-use are reflected in a variety of different and hard to compare and standardise prices.

Mineral products are generally sold on a cost & freight (CFR) or on a cost, insurance & freight (CIF) basis at the port of destination, or on a free-on-board (FOB) basis at the port of origin. Typical sale contract include:

- **Spot or over the counter (OTC) sales** to individual buyers, merchants or on terminal markets.
- **Off-take agreements** of various durations from short-term (typically one to three months) to longer-term (typically lasting one year or more) with prices set with reference to established terminal markets such as LME, LBM, or indices such as Platt's IODEX, various Metal Bulletin's indices, etc. The longer-term agreements significantly reduce the sales volume risk, but generally not the price volatility risk and provide a degree of revenue stability to a project facilitating the securing project finance for its development.
- **Hedging contracts**, in the form of **forwards** directly with customers or **futures** through derivatives markets.

Initially the role of marketing subsidiaries tended to be that of a **sales agent** that is to say of a cost centre, justifying remuneration on a cost plus basis. More recently marketing subsidiaries have sought to become **fully-fledged marketing hubs** adding greater economic value. This may involve taking legal and, less frequently physical, ownership of the mineral product(s) and generally on-selling it to unrelated end users 'on the high seas' typically through so-called 'triangular' transactions as illustrated in Fig. 9.4 (modified from Guj et al. 2017). Further value is added by undertakings a range of specialised activities in the area of marketing, such as engaging customers, handling customers lists and orders, processing of related financial transactions, product support and development of innovative marketing systems and strategies, and of shipping/distribution, such as negotiating and

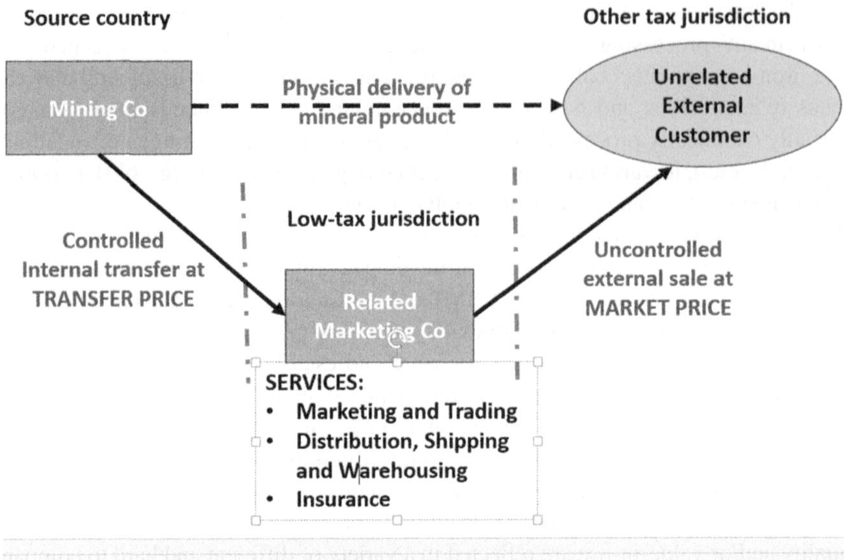

Fig. 9.6 Schematic 'triangular' sale/transfer transaction (modified after Guj et al. 2017)

arranging ship chartering and insurances, possible transshipping, warehousing and packaging of products for shipping and distribution to customers (as shown in Fig. 9.6 below). These services would justify higher forms of remuneration typically associated with the functional profile of a profit centre.

While the legality of establishing off-shore marketing and sales subsidiaries is not under question, significant controversy has arisen as to the degree to which the fees charged by marketing hubs are set at harm-length. The variation in quality of many mineral products affecting their prices combined with the opacity of many mineral markets with a resulting lack of meaningful comparables and the reluctance of MNEs to disclose the terms of their subsequent off-shore sales to third unrelated parties, make the determination of reliable at harm-length transfer prices very difficult for tax authorities to verify and audit.

Long-protracted and very costly court cases relating to the price paid by for the mineral products and the realism of the fees charged for the services of the off-shore marketing subsidiaries of some major mining companies have recently determined that profit was shifted from the country hosting the mining operations to the low-tax jurisdiction in which the marketing hubs reside, resulting in the imposition of multi-million dollar payments in arrear income tax, mineral royalties and related penalties.

As a general rule the difficulty in establishing reliable transfer prices is greater the more specialised and less routine a service is and the greater its reliance on **proprietary intellectual property (IP)** and **intangible assets**. Consequently, opportunity for misuse of transfer pricing are greater in the provision of specialised **technical/engineering and R&D services**, where a significant component of the charges is represented by licences/royalties charged for the use of patented IP and intangibles than in the provision of more routine intra-group **corporate services**.

The services of a related foreign resident technical/engineering and R&D hub are frequently required at both the exploration and particularly at the development and construction phase of a mining project. These may include multimillion dollar engineering, procurement and construction management (EPCM) contracts, also involving relatively long-term secondment of specialised staff to assist the project in the country. Controversies often arise as to the nature and pricing of these services and as to whether the lengthy presence of technical/engineering hub's personnel in the country constitutes the establishment of a **Permanent Establishment (PE)** which would bring into play different taxation provisions.

In the case of routine corporate services a **direct charge** can be applied where the services are similar to those provided by independent parties, i.e. as a comparable uncontrolled price (CUP) or on a Cost Plus basis. Alternatively an **indirect method**, such as the Transactional Profit Split, may be applied where the services cannot readily be identified because incorporated into another transfer as for example, where centrally administered payroll services are charged on the basis of an appropriate **allocation key**, such as the respective number of personnel employed by the different subsidiaries of a MNE.

Although the OECD has been resisting the suggestion and practice of some jurisdictions to take a formulaic approach to the issue, it has to some degree relented on the use of some percentage mark-ups and the adoption of **'safe harbours'** for some of the more routine corporate services.

An alternative approach, better suited for more complex types of transactions, is for industry to negotiate with government and subsequently regularly update so-called **Advance Pricing Agreements (APA)** setting the terms to be used in the determination of relevant transfer prices. While APA once in place work well for both industry and government their structuring and negotiation is an extremely resource and skill intensive as well as time consuming process, in many ways not dissimilar from the conduct of major audit, and for this reason adoption of APAs in mining in developing has been relatively rare to date.

9.2.7 Disclosure, Documentation and Reporting in a Digital World

There is continuing pressure on companies including mining for greater transparency and timely, comprehensive and relevant disclosure in terms of their international tax accounting. Voluntary international initiatives such as the Extractive Industry Transparency Initiative (EITI) and similar national regimes promoted by Canada, Australia and the US Dodd Frank Act for extractives companies listed on the US Security and Exchange Commission (SEC), have progressively been adopted and are now being followed by mandatory reporting requirements such as those introduced by the European Union (EU) for the mining and extractives sectors in their 2013 Accounting Directive 2013/34/EU. The disclosure required by the European Commission is on a project-by-project, government-by-government and country-by-country basis. These measures have now been extended in scope by **BEPS Actions 12–13** which further address the issue of **Mandatory Disclosure Rules and provide guidance** on **Transfer Pricing Documentation** and **Country-by-Country** (CbC) Reporting.

This unprecedented level of tax reform is taking place against a background of rapidly improving communication, IT and data processing capacity which will allow effective enforcement and utilisation of the Exchange of Information (EOI) provisions contained in many DTAs, which in some developing countries had been inhibited by the paucity and ineffectiveness of their electronic information systems. Importantly, the benefits and speed of exchange of information will be increasingly delivered without having to enter into the lengthy process of negotiating a comprehensive bilateral double taxation agreement (DTA) by subscribing to a multilateral Taxation Information Exchange Agreement (TIEA), or by becoming a signatory to the OECD multilateral treaty on mutual administrative assistance in tax matters.

To ensuring better compliance the OECD (Chap. 5 of updated guidelines) recommends that taxpayers should create contemporaneous documentation,, that is to say at the time of the transaction or by the time they file their tax return, restraining taxpayers from developing justifications after the fact, and to evaluate, prior to filing their own compliance with the transfer pricing rules.

Whether, and at what stage contemporaneous documentation should be provided to the tax administration. Provision at the time of tax filing may not be necessary or desirable as it is likely to create an unnecessary cost of compliance for both the company and the tax administration if there is no reason for its review. Provision on request within specified time limits is a more reasonable option if a need for information arises during the tax administration risk assessment or audit phases, if undertaken.

CbC reporting, which came into force on first January 2016, is primarily for the purpose of risk assessment, and limited to MNEs with annual consolidated revenues in the immediate preceding fiscal year of less than €750 million. The CbC disclosure requirements include the provision of information about the MNE's main business activities that include research and development, holding or managing intellectual property, purchasing or procurement, manufacturing or production, sales, marketing or distribution, administrative, management or support services, provision of services to unrelated parties, internal group finance, regulated financial services, insurance, holding shares or other equity instruments and dormant activities.

Needless to say that CbC reporting involves significant compliance costs for industry and potential benefits for developing countries, which may gain more easily access information on the global operations of MNEs active at in their territories. Much of the information produced by all these initiatives may, however, still prove to be too high level to be truly useful in TP audits but may prove an important incentive for greater accountability and disclosure in the future.

9.3 Workable Solutions to Improve Mining Taxation Compliance and Administration

There is little doubt that some of the taxation reforms discussed earlier in this chapter will significantly limit the opportunities and scope for mining MNEs to minimise their tax burden at the consolidated level through deliberately artificial tax strategies and global structures.

Mining MNEs will have the option of either gradually restructuring their global business placing greater emphasis on optimising possible logistic rather than rapidly disappearing tax advantages or to re-direct substance into some of the jurisdictions in which they currently have a low physical presence so as to better justify the level of profit shifted to them. Notwithstanding, the inevitable result will be that the mining industry with time will have to pay more in taxes and that the relevant revenues may be redistributed differently, and hopefully more fairly, among different countries better reflecting where and the degree of value that was added by activities carried out along the mining value chain within their borders. This will inevitably affect shareholders' value and redefine the financial criteria to be applied in determining the commercial feasibility of mining projects.

Due to tax minimisation, and other, primarily environmental reasons, the political climate in a better informed and connected world, has become somewhat antagonistic towards the mining industry and public opinion, fuelled by the intervention of NGOs, strongly supports any government initiative designed to close tax loopholes and to impose draconian penalties on tax evaders. The release in 2016 of over 11 million legal and financial documents relating to over 210,000 offshore companies from the law firm Mossack Fonseca in Panama and the related headlines did not help. All this has inevitably created significant reputational damage and promoted a generally adversarial relationship between the mining industry, the tax authorities and the community even for companies that by and large behaved in a responsible manner, including some larger mining groups with an established track record of regular and voluntary public disclosure of their tax-related policy as part of their annual report or as specific reports.

The time has come to redress the situation. Tax authorities must re-orientate their *modus operandi* from that of a 'policeman' to that of a 'service-provider' attempting to relate to taxpayers as 'customers' and to create and promote a 'culture of compliance'. This will be helped by tax authorities improving the clarity of relevant regulations and related administrative procedures, the prompt provision of required rulings, user-friendly web based documentation and search facilities, readily available and competent advisory services and adoption of effective mediation, conflict resolution and appeal processes designed to avoid as far as possible legal proceedings.

It will also be necessary to ensure that the level of sophistication and complexity of the tax legislation does not exceed the administrative capacity of the agencies that must enforce it. In this context it will be helpful to reduce the red tape and develop a range of simplification measures, such as the already mentioned safe harbours, advance pricing agreements, etc.

Critical to success will also be the need to attract, develop/train and retain appropriately skilled human resources and to develop sophisticated databases and IT systems. In the context of developing countries, of course, this is more easily said than done.

On the other side of the fence, taxpayers should accept the inevitability of having to pay an appropriate level of tax and become voluntarily more open in their dealing with tax authorities, refraining from delaying the release of information required during initial inquiries and/or audits thus avoiding as far as possible potentially lengthy and costly disputes. Companies should proactively seeking constructive engagement with and advice from tax authorities if they are contemplating any change, such as restructuring or expansions of their operations that could significantly affect their tax profile. In fact there is a lot to be gained from improving the amount and tenor of communication with tax authorities and their knowledge of the technical and financial characteristics and challenges central to the success of the company's mining business. There are promising signs of this improvement in the relationship between mining companies and tax authorities already happening in some jurisdictions.

Repairing reputational damage and improving relations with other stakeholders, on the other hand, may prove to be a much harder and long-protracted task. The mining industry has in the past been naïve in believing that people would automatically see the connection between their living standards and the supply of mineral and energy commodities and as a consequence harbour sympathetic feelings towards it and support their development. Redressing this state of affairs will require significant effort at both the individual enterprise level and collectively through the industry's representative bodies to effectively document and publicise both the improving ethical behaviour and the significant contribution of the industry not only to the state revenue but also to other areas of the socio-economic wellbeing of the population at large.

Take Away Points

Government's point of view

- Government should make sure that when formulating sophisticated mineral policy and framing the related regulatory and fiscal legislation and regulations it has or can promptly develop the administrative capacity and systems to enforce and administer them. This is often not the case in many developing countries.
- While regulatory and fiscal legislative amendments may become necessary over time, changes should not be sprung on industry without some previous involvement and consultation while they were formulated, i.e. the principle of 'no surprises' should always be applied. Failing to do so will heighten the perception of the country's sovereign risk and have significant negative consequences in terms of FDI.
- Global restructuring of the mining industry has resulted in many of the

necessary goods and services being provided to the mining operations by service centres or hubs residing abroad mostly in low-tax jurisdictions.

- This has two effects: (1) significant base erosion and profit shifting (BEPS) from the country hosting the mining operations and (2) a very large number of cross-border transactions between related parties at transfer prices not set at arms-length in a contestable market.
- Governments should, if they have not yet done so, consider becoming a signatory to the OECD's BEPS Action Plan, which seeks to put in place measures to prevent excessive tax minimisation by MNEs through global structuring on the basis of 'treaty shopping' and abuse of transfer pricing, particularly relating to the provision of high value goods and services (i.e. financial/treasury, marketing, technical/R&D, etc.) involving hard to value specialised skills and proprietary IP.
- Given its global scope the solution to BEPS requires that all signatory jurisdictions should promptly introduce amendments to their domestic fiscal legislations consistent with the recommended 15 BEPS Actions and put in place the administrative capacity to effectively enforce them.

Mining industry's point of view

- Industry should establish effective channels of communication with the various ministries and government departments involved in its regulation at both the collective (e.g. Chamber of Mines) and individual company level with the objective of being involved, ideally participate in the formulation of policy affecting the sector or their company. Communication should be as far as possible open and constructive and criticisms, where warranted, based on logical arguments. An endeavour should be made to foster a co-operative rather than an adversarial relationship with the government of the country hosting the mining operations.
- MNEs should accept that the BEPS Action Plan initiative is actively being implemented by the vastest majority of jurisdictions and that many previous opportunities for tax minimisation through restructuring and transfer pricing will progressively close with the inevitable consequence of having to pay more tax at the aggregated corporate level.
- In this light MNEs should consider restructuring to achieve greater logistical, operational and marketing synergies rather than decreasing tax minimisation.
- They should also reassess the merit of investing in certain jurisdictions given their rapidly changing fiscal profiles.

References

Guj P, Martin S, Maybee B, Gosai N, Cawood F, Bocoum B, Huibregtse S. (2017) Transfer pricing mining with focus on Africa: a guide book for practitioners. Published jointly by the World Bank, the IM4DC and the CET International Monetary Fund. 2014. Spillovers in international corporate taxation. Policy Paper 2014

IMF (2014) http://www.imf.org/external/pubs/ft/weo/2014/02/

OECD (2015) 2015 Final Reports, OECD/G20 Base Erosion and profit shifting project. OECD Publishing, Paris. http://www.oecd.org/tax//beps-2015-final-reprts.htm
Organization for Economic Cooperation and Development (OECD) (2012) Transfer pricing guidelines for multinational enterprises and tax administrations
Rodik D (2017) Straight talk on trade: ideas for a sane world economy. Quarterly Milken Institute Review. Princeton University Press, Princeton

Chapter 10
Discussion and Ideas

Abstract Every country has its own fiscal regime when it comes to natural resources. Typical components of these regimes include corporate income tax (CIT), royalties, potential free-carried or participative interests, potentially legislated local equity participation, import and export duties, withholding taxes, tariffs, skills development levies, and a number of other tailored imposts. Consequently, the investment attraction of different countries from the point of view of their fiscal regime cannot be compared on the basis of each taxation instrument in isolation but requires consideration of the combined country's 'fiscal package'. Similarly because to the variety of fiscal instruments available, a number of different combination packages may be deemed to be equitable to both the project owner and to the government simultaneously.

Keywords Fiscal regime · Taxation · Royalties · Free carry · Skills

10.1 Current State of Play: Comparing Different Mining Fiscal Regimes

No two countries' fiscal policies are the same and no two countries give the same weight to their various economic objectives when it comes to their fiscal regimes. That said, commonality exists in so far that each country endeavours to ensure adequate and, as far as possible, stable revenue is received by the State irrespective of their preferred combination of available income-generating taxing instruments in order to achieve economic growth in a sustainable manner.

Although numerous countries are blessed with an abundance of natural resources, it is only through active exploration activities that it may be determined whether these natural resources may be economically exploitable or not. The challenge then becomes to attract local or, more often, foreign investment capital to develop these resources by way of economic incentives and creating a perception of political and fiscal stability.

E. Lilford, P. Guj, *Mining Taxation*, Modern Approaches in Solid Earth Sciences 18,
https://doi.org/10.1007/978-3-030-49821-4_10

Many mineral-rich countries have generally been exposed to mining activities over the years and have established and gradually refined appropriate fiscal regimes to govern their mining sector. Jurisdictions that have demonstrated economic success often tend to be those that have used some of the incoming monetary benefits from mining activities to at least sustain, but typically upgrade national and local infrastructure and demonstrably improve social welfare in terms of delivery of health, education and municipal services. Economic success becomes self-sustaining when the perception of sovereign risks, which initially inhibited a foreign investment attraction, diminishes to the extent that existing and new investors plough capital into the country, not limiting their investment to the natural resources sector.

As already discussed, a typical mining fiscal regime adopted by a government to capture economic rent from its mining industry will include a number of common taxing instruments and possibly a few other unique components of the package not strictly classifiable as taxes. Typically common tax and non-tax components include:

- corporate income tax based on the entity taxable income;
- payroll/employees' tax;
- a mineral royalty, with the type of royalty and rate dependent on the commodity type;
- import/export duties, allowing for negotiated moratorium over the initial, capital-intensive project construction period and/or during ramp-up;
- environmental rehabilitation contributions for environmental reparations after closure;
- capital gains tax on disposal of assets or for change in ownership;
- withholding taxes;
- value added or goods and services tax (VAT or GST); and
- fringe benefit tax (FBT).

In addition to the above sources of State revenue, common sources of additional, albeit lesser, income, benefits or participation may include:

- an imposed skills development tax or levy, to be used solely for the benefit of employees or their immediate beneficiaries to equip them with additional or diversified skills;
- adult education, providing basic skills training to employees and their direct families who may have missed or been deprived of a basic education previously;
- participative or free carried equity interest at a government level or a wholly-owned government subsidiary;
- participative equity interest provided on a subsidised repayment basis for local participation by the community and/or employees;
- infrastructural development requirements, positively impacting communities and regions (e.g. clinic or hospital, schools, roads, rail networks, potable water supply, electrification, housing, recreational amenities, etc.); and
- local procurement requirements for goods and services.

Table 10.1 shows the main components of the fiscal regime 'package' of selected countries as it relates to their mining industry. The percentage royalty rates listed in

Table 10.1 Comparison of selected mining fiscal regimes *(royalty rates are ad valorem unless otherwise stated)*

Comparison of Selected Mining Taxation Packages

NOTE: Royalty rates are ad valorem unless otherwise specified. Specific weight-based royalties, that may co-exist with the ad valorem systems not listed.

Country	CIT rate	Royalty range	VAT/ GST	Import/ export duties	Withholding tax	Govt./local/social participation
Argentina	35.0%	3%	n/a	5–10%	0%	Y
Australia	30.0% being progressively reduced to 26%	All states 1.6–15.0%	n/a	n/a	30%	N (Aboriginal rights)
(a) Western Australia		2.5–7.5%				
(b) Queensland		1.5–15.0%				
(c) New South Wales		4.0–8.2%				
(d) Victoria		2.75%				
(e) Northern Territory		22.5% of profit				
(f) Tasmania		Hybrid, 1.9% of NSR + profit. Max 5.35% NSR				
Botswana	22.0%	3–10%	12%	Variable	7.5%	N
Burkina-Faso	28%	5%	18%	30%	12.5%	Y
Brazil	34.0%	1–3.5%	3.65–9.25%	0%	0%	N
Cambodia	20%	3–5%	10%	Variable	0%	Y
Canada Federal (excludes Provinces)	15.0% (Fed. only)	1–20% of income/profit	n/a	n/a	25%	Y
(a) Ontario	10%	10% of profits (5% for remote area)				
(b) Saskatchewan	10%	5% of income to 1Moz or 1Mt base metals, 10% above it				

(continued)

Table 10.1 (continued)

Comparison of Selected Mining Taxation Packages

NOTE: Royalty rates are ad valorem unless otherwise specified. Specific weight-based royalties, that may co-exist with the ad valorem systems not listed.

Country	CIT rate	Royalty range	VAT/ GST	Import/ export duties	Withholding tax	Govt./local/social participation
(c) British Columbia	12%	2% of profit and 13% on revenue				
(d) Quebec	11.6%	16% of profits				
(e) Newfoundland	15%	20% of profits				
(f) Alberta	11.5%	1% of income at mine mouth plus 12% of profits				
Chile	18.5%	0–14% of profit	n/a	n/a	35%	N
China	Nat. 25.0% Prov. 3%	0.5–4%	13%	Variable	10%	Y
DRCongo	30.0% and 50% on super-profit	0.5–10%	16%	n/a	10%	Y
ROCongo	30.0%	3–5%	0.2–1%	n/a	20%	Y
Germany	30.0%	n/a	n/a	n/a	26.375%	N
Ghana	35.0%	5%	15%	n/a	8%	Y
India	25.0%	2–10%	1–5%	10–40%	0%	Y
Indonesia	25.0%	3–7% 13.5% for coal (rates vary with production tonnages)	n/a	n/a	20%	Y
Japan	41%	0.7–1%	10%	n/a	20%	N
Kazakhstan	20.0%	0–5.7%	variable %	Coal 2%	15%	Y
Kenya	37.5%	5–12%	16%	0–5% on minerals	10%	Y
Korea	20%	n/a	10%	n/a	27.9%	N

Laos	20%	2–5%	10%	Variable	5–20%	N
Liberia	30%	3% plus 2% presumptive royalty	10%	2.5–20%	5%	Y
Malaysia	25%	5%	10%	0–10%	15%	N
Mali	30%	6%	18%	Variable	10%	Y
Myanmar	25%	3–8%	10%	n/a	0%	Y
Mexico	30.0%	n/a	16%	n/a	10%	Y
Mongolia	10–25%	Domestic 2.5% Exports 5% + surtax by commodity, price and product type	10%	0–40% mostly 5%	10–20%	N
Mozambiques	32%	1.5–8%	17%	n/a	20%	Y
Namibia	37.5% (55% diamonds)	2–5%	15%	Variable	10%	Y
Nigeria	30%	3–5%	5%	Variable	10%	Not specified
Papua and New Guinea	30% (Non-res. 40%)	2%	10%	Variable	Int. 15% Div. 10%	N
Peru	29.5%	Profit-based at 3 Gov. levels: Min Royalty 1–12%, Special min tax 2–8.4%, Min Contribution 3.4–13.2%	n/a	n/a	5%	Y
Philippines	30%	2–8%	0–12%	n/a	12%	Y
Russia	35.5%	4.8–8%	n/a	Select products	15%	Y
South Africa	28.0%	0.5–7%	15%	n/a	10%	Y
Tanzania	Gen. 30.0% Min. 35%	6%	18%	n/a	10%	Y
Thailand	20%	2–10%	7%	Variable	15%	Not specified
Ukraine	18.0%	Variable/t	20%	Select products	15%	Y
UK	19.0%	n/a	variable/t	Select products	0%	N

(continued)

Table 10.1 (continued)

Comparison of Selected Mining Taxation Packages

NOTE: Royalty rates are ad valorem unless otherwise specified. Specific weight-based royalties, that may co-exist with the ad valorem systems not listed.

Country	CIT rate	Royalty range	VAT/ GST	Import/ export duties	Withholding tax	Govt./local/social participation
USA	21% reduced from 48.0%	0–12.5%	4.4%	n/a	30%	N
(a) Montana		5–8% on NSR basis				
(b) Alaska		3–7% of income				
(c) Arizona		2% on income				
(d) California		At least 10% of profit				
(e) Idaho		5% of income				
(f) New Mexico		At least 5% of gross returns				
(g) Oregon		5% of income				
(h) Utah		4% (non-metal) to 8% (metals) of profit				
Vietnam	20% (32–50% precious metals)	1–5%	10%	0–45%	10%	N
Zambia	30%	5–10%	16%	0–25%	20%	Y
Zimbabwe	25% (+3% AIDS levy)	1–15%	15%	Variable	10–15%	Y

Table 10.1 are *ad valorem* unless otherwise specified. They are levied on a range of royalty value bases spanning from the gross value of metal to its net value at the mine mouth, with most using its fob value at the port of export. Table 10.1 does not include a range of less significant specific, generally weight-based, royalties that are normally applied to lower-value minerals and often co-exist with the listed *ad valorem* systems. In general the trend has been for developed countries to reduce their CIT tax rates to stimulate their economies, notable examples are the USA which recently has reduced its tax rates from 48% to 21%, the UK to 19% and further to 17% in Q2 2020 and Australia which will progressively reduce its tax rate from 30% to 27.5% and further to 25% by 2022 (26% by 2021). By contrast some developed countries, such as Japan at 41% have maintained very high CIT rates.

In the developing world, applicable CIT rates typically range between 20% and 35%.

10.2 Major Trends in Mining Taxation Policy and Administration Development

10.2.1 Mineral Production and Related Government Revenues Will Continue to Grow Because of Growth in Population and Per Capita Consumption

The demand for mineral commodities can be confidently expected to continue to increase over time for two fundamental reasons, being:

- an increasing global population; and
- an expected increase in demand per capita (intensity of use) approaching the absolute market potential (absolute market potential is a term used in Economics to describe the highest possible demand rate if each user uses that commodity to its full potential use on each occasion it is used.

Of course the increase in demand will not be uniform across all mineral commodities because of the different requirements of evolving technological and socio-economic needs and expectations. Already a challenge is emerging in ensuring the sustainability of supply of a range of critical raw materials (CRMs) in rapidly increasing demand generated by their use in the broad area of clean energy applications.

An example of rapidly increasing intensity of use in developing or emerging economies is provided by copper. Over a decade ago, the average annual copper consumption rate for developed economies was around 10.0 kg per capita. At the same time, China's consumption rate for copper was around 3.0 kg per capita (slightly over 4 Mt per annum). If Absolute Market Potential provides guidance and if China's demand increases to approach the average consumptive rate for developed countries of 10.0 Mt per capita, that is to say an increase of 7 kg for each of the approximate 1.5 billion people of China, the potential demand for copper may increase by almost 10 Mt of copper per annum equivalent to around 80% of current global copper production.

Similarly India's current per capita copper consumption rate is a mere 0.5 kg (www.https://economictimes.indiatimes.com). With the same Absolute Market Potential of 10.0 kg per capita, there may be significant additional demand from India for copper and other commodities.

While increasing demand will also stimulate recycling, for many commodities the bulk of the demand increase will probably be largely satisfied through additional mining output.

As this type of logic can be extended to the demand for most mineral commodities, it is realistic to expect that mining will remain a necessary activity in the future and therefore it is important for both governments and producers to formulate robust and enduring policies to manage the industry and equitably share in the associated benefits in the future.

The expected global population increase will not put pressure exclusively on the demand for minerals and metals, but also create competition for other land uses. Sound economic comparisons should become increasingly important to determine the best use of the available land, often in competition between agricultural, residential, industrial and natural resources exploitation, assuming the land in question is mineralised. Comparisons should be based on fully fledged cost-benefit analyses attributing an opportunity cost for the next best land use foregone in favour of mining which should inform the formulation and implementation of a well thought out and economically efficient mining fiscal regime.

Importantly, given the finite life of mining operations and due consideration to the fact that natural resources properties, once exhausted, are typically rehabilitated and returned to the state, it is more appropriate to cast the dialogue in terms of successive rather than alternative land uses.

The relevant government will then have the opportunity to receive income from mining taxes and royalties and, once mining operations have discontinued due to ore depletion, thereafter taxes from farming or other economic activity taking place on that land.

10.2.2 Continuous Change and Instability in the Fiscal Regimes of Developing Countries

Key to a successful mining economy is that both governments and producers must play their part and bear their proportional shares of the costs involved in its progressive establishment and development. In this context, the "costs" to be borne by governments do not generally involve actual financial contributions out of existing state funds but rather to a willingness to temporarily forego or defer future revenues by granting specific concessions or reprieves during the early and critical stages of a mining development. These concessions may be import/export duty and VAT moratorium periods, royalty and/or corporate income tax deferments during a project's ramp-up to full production, etc.

Contrary to the above, many countries, and notably emerging market economies, have been attempting to vary their fiscal regimes in order to secure a larger share of or accelerate the reception of the economic benefits arising from the exploitation of their natural resources. Examples of this behaviour within the last 10 years include:

- In late 2019, **Mali** announced that mining companies will no longer be exempt from VAT during production. Additionally, the country's previously stated stability period will be shortened considerably from the previous 30 years. These changes to their mining fiscal regime are likely to have notable impacts on future investments into that country's natural resources. It can be expected that investors will react by planning in a manner that achieves the bulk of their investment returns within the stability period, irrespective of whether the resulting mine planning in terms of annual output, cut-off grades and size/life of mine may lead to suboptimal exploitation of the resource. Risk management would justify high-grading the deposit in the shorter term and consequent sterilisation of some of the resources because of the lower weighting given to potential returns that may have been achievable in the absence of the limitation in the stability agreement.
- **Zambia** introduced and then rescinded its Super Profits Tax. Also, in an endeavour to eliminate tax evasion, declared all mining contracts null and void, abolished tax on income relating to mining and virtually doubled the royalty rates, generating significant industry reaction and a political crisis that culminated with the virtual re-establishment of the status quo;
- the **Democratic Republic of Congo** increased the royalty rate for 'strategic minerals' from the general 2% rate to 10% and included cobalt among the strategic minerals. It also introduced a 50% tax on super-profit, defined as taxable income realized when commodity prices rise 25% above levels included in a project's bankable-feasibility study. These changes are intended to apply immediately to new and existing projects.
- **South Africa** drafted their Minerals Charter III that increases the equity participation of local communities and employees. Charter II requires broad-based black economic empowerment (B-BBEE) participation at 26% whereas Charter III increases this overall B-BBEE participation to 30%. Of this revised 30%, 5% will be free-carried by employees and 5% free-carried by the local community, neither of which free-carry was contained in Charter II;
- **Zimbabwe** called for a 51% indigenisation requirement, but has now refined the policy to exclude many minerals;
- **Australia** introduced an MRRT on iron ore, coal and on-shore petroleum production, and then rescinded its application to the first two commodities;
- the **United States of America** progressed the introduction of tariffs against Chinese goods, notably mineral products including aluminium and steel;

- **Ghana** increased corporate income tax rate from the generally applicable 30% selectively to 35% for mining;
- Between 2015 and 2018 **Tanzania** expedited through parliament without consultation with the mining industry, key tax changes including an increase in the royalty rate from 4% to 6%; reduction in depreciation rate for mining related assets from 100% to 20%, i.e. on a straight line over 5 years; set a minimum non-dilutive government free carry interest at 16%, with potential to lift it up to 50%. It also alleged that a mining company engaged in deliberate misinvoicing grossly understating the grade, hence value, of its exports of mineral concentrates to a related foreign company. This culminated in a suspension of concentrate export and in a long and as-yet unresolved dispute, which, together with the hike in taxes, has negatively affected Tanzania as a desirable investment destination.
- **Indonesia** banning export of nickeliferous laterite ore and of some other unprocessed ores in an effort to stimulate a higher rate of investment in domestic mineral processing.

10.2.3 Trends Towards Greater Harmonization of International Mining Tax Regimes

Fiscal reform in recent years has nonetheless displayed a tendency for CIT and royalty rates to be amended to within more narrow ranges than in the past, with the majority of the former falling mostly between 25 and 35% and the latter between 2.5 and 5%. This trend begs the question as to whether it would be economically feasible to further standardise CIT and royalty rates and the various other components of every country's natural resources' fiscal regime, at the limit having each regime hosting mining projects applying the same charges and rates.

The obvious answer to the above question is a negative one as, having different, unique regimes somewhat defines the concept of "sovereignty" and the attractiveness of a fiscal regime should be inversely proportional to the different level of risk in investing in various countries. In addition, many countries exhibit a competitive advantage for specific commodities over other countries and hence contend that their specific fiscal regime must suit the specific characteristics of their natural resources endowment.

The highest level of commonality should be for each and every country to be completely transparent, consistent and stable in their respective policies and regimes in order to provide a degree of certainty to potential local and foreign investors. While each country can exhibit uniqueness in their policies and associated regulatory regimes, policy-makers must bear in mind that each country is also competing for a finite pool of investment capital from the wider world and that countries offering potential investors an attractive investment proposition will ultimately attract the lion share of that hard-to-find investment capital.

10.2.4 Continuing Pressure for Greater Government and Local Participation in Mining Projects and for the Establishment of Downstream Processing Facilities in the Developing Countries Hosting the Mining Operations

Since mining is a finite activity due to the depleting nature of an orebody being exploited, it is in the interests of any country that hosts a mining operations to receive its fair share of the economic benefits associated with the exploitation of that orebody. As can be expected, the mine owner/operator and the government have different views of what a 'reasonable' fair share may be. In light of competition amongst different countries to attract FDI into their respective resources industries, the requirements surrounding government and local participation in mining operations as well as the requirement for companies to establish downstream processing infrastructure in-country ,may prove to be an inhibiting factor.

Previous chapters in this book have detailed the implications associated with government and/or local equity participation on either a free-carried or on a participative basis. For the mine owner or mine investor, these impositions tend to be economically unattractive and may result in the investment being avoided altogether. From the government's perspective, the reasoning for imposing an equity participation may be due to:

- the accruing of pro-rata benefits by being an equity holder (dividends and other financially beneficial distributions);
- the resulting economic upliftment that may be expected arising from the returns passed on to the government or community;
- the associated upskilling associated with the government shareholders and their representatives who will need to develop new and more sophisticated skills in order to fully understand the implications of their equity positions.

Other than the deemed benefits of holding equity, governments may introduce legislation that encourages or even imposes on mining companies the obligation to establish downstream processing facilities in the country. Alternatively, a resources-rich country may include specific, condition designed to dissuade producers from selling under- or unbeneficiated mineral product to offshore processing facilities, making it more economically attractive to establish in-country processing.

In the final analysis investment in downstream processing should be commercially justifiable in its own rights or by virtue of subsidies and other incentives provided by government as for instance by offering cheaper electricity, industrial land, reduced taxes and other economically attractive incentives. These incentives may be packaged into the establishment of a special development zone (SDZ) wherein mineral processing may be afforded a range of cheaper inputs.

Asides from the direct and indirect benefits derived from downstream processing in their countries governments will continue to look more closely at ways of securing increased economic benefits by looking at ways to enforce skills-transfer and skills upliftment and development for their citizens, which would result in an overall betterment of their country's wider economy over time beyond just the mining sector. As already discussed, a common fiscal incentive for industry to develop downstream processing capacity may be in the form of introducing scaled royalties such that the less a material is processed, the higher the associated royalty rate,

Intuitively, as orebodies are depleted around the world, new projects are being developed to replace the depleting resources and maintain or ideally expand the resources endowment of the various commodities. In a number of cases, the depletion of a specific orebody may not mean that the related processing facility should be dismantled or removed, but that the owner should look for alternative sources of feed material from other producers or by precipitating the development of new resource projects within a radius of potentially economic transport cost in the same or in an adjoining jurisdiction. The imposition of governments regulations attempting to motivate producers to build new processing infrastructure in the country may, in the absence of a business case, represent a significant deterrent for investment at all or, if the processing facilities are made part of a single integrated project, they may lend being cross-subsidised by the upstream mining component of the business, which would be a suboptimal investment from the company's point of view but may prove attractive from the government's point of view when all the socio-economic spin-offs and positive externalities are taken into consideration.

To attract foreign investment, some governments are often prepared to negotiate moratorium periods during which no import or export duties or mineral royalties are paid. While this is certainly an attractive incentive for mining companies, it can, as already discussed, be detrimental to the government if the incentives are not clearly specified creating an opportunity for the mining company to high-grade the mineral deposit or expand production during the moratorium period, resulting in a lower average mining grade and residual mining life respectively once the moratorium period ends. This will impact later cash flows, returns and value and consequently reduce the overall potential revenues to the government.

10.2.5 Tightening of International Tax Conventions and Rules to Combat Base Erosion and Profit Shifting (BEPS) and Increased Use of IT for Inter-Jurisdictional Tax Information Exchanges in Support of Audits

At the time of writing in excess of 100 jurisdictions had indicated their intention to implement the 15 BEPS recommendations by introducing appropriate amendments to their domestic tax legislation and related regulations. Inevitably these legislative processes will take some time as they are progressively implemented and their enforcement in practice will in some cases present some teething problems. The result will be a period of continuous changes and progressive tightening in fiscal rules requiring industry updating of their accounting tax systems to comply particularly in the context of transfer price documentation and reporting.

Until recently tax authorities had significant difficulty in obtaining necessary auditing information relating to the foreign jurisdiction component of taxpayers' cross-border transactions. This provided the opportunity for taxpayers to frustrate or delay auditing processes to a point where they became counterproductive for the tax authorities to continue to pursue. The inclusion of exchange of information clauses in double taxation agreement, although cumbersome to utilise, helped somewhat but only to the limited extent of nations with which agreements were in place. Now with the introduction of BEPS multilateral instruments the process of exchange of information between different tax jurisdictions will be significantly expedited and any progressive BEPS fiscal initiative automatically and immediately implemented by the various jurisdictions.

10.2.6 Fiscal Challenges Created by the Increased Provision of Digitally Delivered Services Between Related Parties of MNEs

Technological and communication innovation has created the opportunity for mining MNEs to provide a large number of services to their mining subsidiaries using digital technology. As a consequence, the service hubs providing a range of both routine (corporate services) and specialised services (engineering, R&D, legal) and related high value intangible assets can be located anywhere in the world. The current strong trend towards automating many traditional mining operation functions further increases the opportunities to separate them physically from the mine site. As discussed, all this has profound implications as to the allocation of profits to the various entities and jurisdictions involved.

In early 2019 the OECD, with endorsement from both the G20 and the G7, has initiated a consultative process, commonly referred to as BEPS 2.0, designed to address the tax challenges of the digital economy and ultimately revise existing profit allocation and nexus rules and developing new global minimum tax rules. A deadline to achieve international agreement to implement these changes, which will have a significant impact on MNEs, has been set for 2020.

The first pillar of the OECD program is to redistribute some of the taxing rights to the market jurisdictions that is to say to the jurisdictions hosting the mining operations which are the major recipients of the digital services even though the relevant activities were not conducted there. The second pillar seeks to establish minimum levels of taxation as an anti-base-erosion mechanism.

While it is premature to speculate as to what the final guidance will be, the clear intention of increasing the proportion of profits which will be taxable in mining countries, will need to be seriously considered by mining MNEs and will be another factor in determining their future restructuring and geographical deployment of their resources. Once again MNEs will need to assess the benefits from any logistical and operational synergy which could be derived by necessary restructuring to offset the inevitable increase in their tax liabilities at the consolidated level.

10.2.7 Greater Community Involvement in Mining Tax Issues Affecting the Licence to Operate and Mining Companies' Reputational Damage

One of the major stakeholders in any mining operation is the local community and any other regional community that may be directly affected by the development and exploitation of a specific natural resource. Many emerging economies have introduced regulations that seek to accommodate these communities through either work opportunities, skills development programmes, economic benefits to the community through direct income, dividends and economic upliftment, wealth participation through equity participation and local infrastructure benefits, etc. Nonetheless these initiatives may not necessarily provide an adequate impetus for a company securing of its social license to operate.

A social license to operate is of course not an actual formal license granted by authority as a condition for a specific tenure. It represents an unwritten agreement between a mining company and an affected community proffering that the mining company will work together with the community on a best endeavours basis to ensure that the community is, on balance, not negatively impacted at by the mining activities. These social license aspects often include, but are not limited to:

- provision by the mining company to the community of the opportunities to be heard due to mining and associated activities taking place in close proximity to their dwellings, farmlands, recreational settings or other important areas;
- opportunities provided to the community to be employed, upskilled, provided with necessary facilities and services, etc. consistent with their desired development;
- ensuring that environmental, cultural and social areas (including heritage and burial sites) are not impacted negatively through dust, noise, smell or visibility arising from activities; and
- improving, or at least not negatively impacting, the overall well-being of the community through the support of the mining company, with either financial or in-kind assistance, including possibly school or tertiary education bursaries and trades apprenticeships, as well as providing entrepreneurial skills and opportunities.

Some communities may become the direct recipients of royalties or tax benefits arising from mining activities. While the actual receipt of pecuniary benefits may not be ideal for all communities as has been witnessed in some parts of Australia, success stories are evidenced when looking at the development of the Royal Bafokeng Nation in South Africa, which receives royalties from specific platinum mining operations, and to a lesser extent the Bakgatla-ba-Kgafela people in the Pilanesburg region of the same country, benefiting from different platinum mining royalties.

For the communities to be benefited in some way or another, they must be informed of the mining operation's development and the ways in which they can participate to its success and derive benefits. As can be expected, without some community's buy-in, an existing or proposed mining operation is destined to fail at some level due to the political power the communities can wield if they feel disenfranchised or disengaged. The power they hold is associated with their capacity to withhold this virtual social license to operate. In practice, even the government or associated authorities cannot override active community's disapproval and related NGOs' pressure or, if they can, do so at their own peril. Mining companies not only carry the risk of being unable to operate effectively without the endorsement of the local community, but they also run the risk of a heightened reputational damage if they do not adequately engage with the community to redress any unacceptable situation.

10.2.8 Growing Impetus to Include Taxation of Carbon-Emission by Producers

There is an increasing voluntary effort by some mining companies to reduce their carbon emissions throughout their mineral extraction and processing activities in an endeavour to mitigate climate change effects. However, voluntary action alone may not be sufficient and consequently there is increasing focus by governments to impose some kind of tax on large carbon emitters in their jurisdictions. The effect will be to put a specific value on carbon which may be different in each country, but in all likelihood determined with cognisance of the carbon values/prices adopted in other international jurisdictions.

Once a carbon tax is believed to be appropriate, governments will likely introduce a system to determine how many carbon credits each carbon emitting enterprise is entitled to and procedures through which lower carbon-intensive producers, through cleaner technologies including those associated with sequestering carbon, will be allowed to trade (sell) their excess carbon permits and therefore receive a financial benefit for being a cleaner producer.

The counter-side will be that higher carbon-polluting operations and companies will have to buy additional carbon permits to allow them to continue producing at high carbon levels. The cost of these permits will add to their operating expenses therefore creating an economic incentive to become.

The government will control the number of "polluting" permits awarded to various companies, at a set cost, which means that the government will receive an income (a carbon tax) from the sale of the permits. Once permits are owned by an entity, it will be free to trade them and in the process determine a fair price for carbon. Eventually international trading in carbon credits will with time homogenise carbon prices across national boundaries.

Take Away Points

Government's point of view

- Since mineral resources are a national asset, an appropriate fiscal regime relating to the minerals industry is required by each country. The regime captures one or more of CIT rates, royalty rates, free carry or participative interests, various duties on exports and imports, withholding taxes, tariffs, skills development levies and other imposts.
- To retain existing and to attract additional foreign direct investment, the fiscal regime applied to the resources sector must be deemed to be fair by the mining industry. Governments must balance the various components to ensure this is the case.
- Different countries will derive a fiscal regime balance dependent on a number of factors, not least of which is the importance of the mining sector to their overall economy.

Mining industry's point of view

- Companies can choose where to invest, both from a commodities point of view as well as a country point of view. Where countries host similar commodities, the country that offers the propensity for the greatest investment returns commensurate with sovereign and other risks, will be the preferred investment destination.
- A clear, stable fiscal regime is desired by Companies that are likely to invest hundreds of millions of dollars into a new mining project. Sovereign risk is critically assessed prior to investing.

Chapter 11
Ideas About the Way Forward

Abstract There is no perfect solution to a fiscal regime that will fully satisfy a government and industry simultaneously. Compromises need to be made and agreed upon. Governments wield the authority to render changes to attract foreign investment while industry has access to development capital. An equitable regime will attract foreign investment and will ensure that the government receives its fair dues from resources exploitation.

Keywords Fiscal regime · Policy · BEPS

11.1 General

It is recognised within governments and businesses alike that natural resources occurrences within a country are finite in nature, that is to say they will deplete over time as they are progressively exploited and that, being a national asset, their exploitation must provide benefits not only to investors but to the country as a whole. There should also be general agreement that there needs to be a balance between the benefits that may accrue to the operating company and the benefits that may accrue to the government over the period during which the mineral resources are exploited. In the case of mining companies, the benefits should be financial and measurable in terms of their adequacy in attracting and retaining foreign direct investment in the country making allowance for the country risk. The benefits to the government and hence the wider community, by contrast, should be socio-economic and directed to foster the country's sustainable development. The benefits accruing to the country may be measured in a number of ways, including:

- Direct financial returns, derived through mechanisms including a variety of taxes, mineral royalties, import/export duties, tariffs, dividends on carried or free carried equity interests in mining projects, etc.; and
- Indirect monetary and non-monetary benefits, derived through mechanisms including local and broader employment multipliers, subsidised local equity participation, skills transfer programmes, education programmes, provision of

E. Lilford, P. Guj, *Mining Taxation*, Modern Approaches in Solid Earth Sciences 18,
https://doi.org/10.1007/978-3-030-49821-4_11

specific infrastructure (hospitals/clinics, schools, roads, other infrastructure) in excess of project requirements, entrepreneurial programmes, establishment of secondary industry to accommodate local procurement, diversifying services offered by main-stream suppliers to new, establishing or growing service providers, etc.

The above principles are well established and by and large accepted in both government and industry circles. They also form the basis on which community, NGOs and other stakeholders' expectations are formed and of their power to influence the political processes of mineral-rich countries. This is a good start, but what can government do to further improve their fiscal policy formulation and administration systems and how can industry help in the process thus leading to more balanced and equitable economic outcomes, a more harmonious, constructive and synergistic relationship leading to less disputes and overall less onerous compliance costs?

A primary consideration requiring attention when a government is setting, or ill-advisedly revising, its various taxation, royalty and other fiscal rates is to ensure that the related imposts do not become so punitive as to scare away foreign investment, particularly in view of the fact that most local, emerging economies cannot generate enough domestic capital and skill bases to independently develop their own natural resources base. The attraction and retention of foreign investment is crucial for the development of not only natural resources projects themselves, but also of the numerous peripheral infrastructural developments necessary to ensure a project is economically viable and that the product can reach the intended, typically off-shore market.

One would imagine that, given the economic importance of some of these policy decisions, they should be informed by rigorous economic modelling and evaluation and their formulation supported by wide and meaningful consultation between government and the mining industry and, as necessary, with other stakeholders. This is, however, seldom the case as in the majority of cases the relationship between tax authorities and the mining industry has been cautious, clouded by mutual suspicion, if not outright adversarial.

The tenor of the relationship between government and industry has not been helped by the latter which, driven by its objective of maximising shareholders' value, has pushed the tax-minimisation principle of paying 'the least and the latest' tax to the extreme by exploiting, within the bounds of the law, every possible loophole and weakness in the tax legislation. Artificial tax minimisation and even evasion schemes have also been facilitated by the general inability of tax authorities, particularly in developing countries, to effectively administer their tax regimes and carry out timely audits because of lack of adequate resources and/or of specialised skills.

These problems have been exacerbated by increased globalisation of the mining industry with resultant fragmentation of the mining supply chain and exponential increase in cross-border transactions between related parties all members of the same multinational enterprise (MNE) for the provision of a variety of goods and services on the basis of 'transfer prices' rather than prices set in a contestable market between unrelated third parties. MNEs have adopted corporate structures designed to minimise their tax payments at the consolidated level by establishing service centres or hubs for the provision of specialised and valuable goods and services in foreign, generally low or no tax jurisdictions, thus eroding the tax base of and shifting profits from the country hosting the physical mining operations.

11.2 What Government Could Do

11.2.1 Promoting Growth of the Mining Industry by Supporting Mineral Exploration

An alternative to securing a larger share of the rents generated by the mining industry is for government to do all it can to expand the size and scope of the mining activities carried out in the country. In other words, if the current level of taxation is considered in line with a jurisdiction's competitive capacity to attract international investment, government should frame its policy and strategies to pursue a **constant proportion of a growing pie**, rather than aiming for a growing proportion of a static or even shrinking one. The future economic importance of the mining industry in a country can be influenced by a range of strategies including:

- **Facilitating expansion of current production** through simplification and streamlining of relevant approval processes and other regulatory constraints. This strategy will be effective as long as an adequate amount of additional resources/reserves are delineated and developed ahead of each year's production. In the long-run, however, this strategy will be unsustainable given the finite nature of the various orebodies and will lead to accelerated exhaustion and progressive closure of existing mines.
- **Promoting and supporting mineral exploration**. Exploration should be facilitated on two fronts by the collection and free dissemination by government of relevant, pre-competitive geoscientific information at a regional scale and timely collection and release to industry of companies' exploration data and reports. Firstly production of the mineral commodities currently being produced should be maintained and ideally expanded through the discovery of new orebodies through both brownfield and greenfield exploration to progressively replace existing mines as they are exhausted. Secondly a concerted effort should be made to identify the potential for the discovery and development of new mineral commodities as yet not mined in the country.

11.2.2 Understand and Communicate Better with Industry

The formulation of effective mineral fiscal policy and related legal and administrative rules and regulations need to be based on a deep understanding on the side of the policy-makers of the technical, operational and financial characteristics of the mining industry and of what makes it flower. An adequate degree of industry knowledge generally exists if the Ministry for Mines and related Mines Departments are charged with the administration and collection of mineral royalties in parallel with a broad range of other industry regulatory functions, such as operational health and safety, compliance with environmental conditions etc., which require frequent interaction and flow of information between industry and government. Effective mechanisms for regular communication in most cases already exist between the technical ranks of Department of Mines and industry both at the individual company level and/or through the involvement of industry representative and lobbying bodies (e.g. Chamber of Mines), but they could generally be strengthened where it comes to matters relating to fiscal policy formulation and administration. These consultative processes build a degree of trust and, to the extent that administration of mineral royalties is closer to the mining operations and less complex and controversial than that of CIT, the relationship between the mining industry and the Mines Departments tends to be less adversarial and more cooperative than that with the tax authorities. By contrast, the rapport between the mining industry and the Ministry of Finance and related taxing authorities, such as the Taxation Office and Customs and Excises, does not involve a significant level of technical communication and interaction and, as a consequence, tax authorities have a lower level of knowledge and understanding of the mining industry and their relationship with industry is generally more formal, remote and less trusting. Industry often complains that government tends to formulate and enforce new fiscal rules and regulations with no or inadequate levels of prior communication and consultation. This can be very detrimental to their relationship and reinforce adversarial attitudes. As discussed in Sect. 11.2.6 below, there is significant scope for improvement in this area.

11.2.3 Formulate and Implement Long-Term Mineral Resources Policy and Plans and Adopt Mining Fiscal Regimes that Balance Short and Long-Term Funding Needs

The government holds all the aces. It is primarily through their policies and regulations that foreign companies will or will not invest in their country. Since mining investments are relatively risky and tend to be long-term (greater than 10 years), companies must be comfortable that they can attract up-front equity capital, with debt to potentially follow once the equity has been drawn down, at a later stage, to develop these specific mining assets. The most important factor associated with this is security of tenure, followed by other aspects of sovereign stability and the associated stability of various policies and regulations.

It is understood that governments would like to start receiving financial benefits from mining, through taxes, royalties, participation, duties, etc., as soon as possible after mining commences. The main reason behind this early-income drive seems to be because governments are chronically short of funds and the politicians and their associated officials tend to focus primarily on their short-term political cycle deploying funds in a manner that maximises their chances of being re-elected beyond the current term. They generally want to be recognised for their achievements while in office and not necessarily set up the foundations for benefits and recognition that may be realised by their successors. This is unfortunate because a typical mining operation will significantly outlast even the longest-serving politician. The community and national interest, however, would, in the majority of cases be better served if the government-of-the-day focussed on the longer-term benefits that could potentially arise from properly planned and executed mining activities and not just on short-term gains.

The development of mining assets often provides the impetus for the establishment of common use infrastructure. These include the general development of:

- pre-developed and services dedicated industrial land estates;
- roadways, railways and associated infrastructure;
- ports, storage and loading facilities;
- local and regional electricity generation and reticulation systems that may be shared with the local community;
- industrial and potable water systems, including possible reverse osmosis desalination plants in arid regions;
- sewerage and other waste disposal systems; and
- the need for a range of services that may be provided by local communities and businesses.

The establishment of these types of common-use infrastructure can be a powerful magnet in attracting FDI and expediting the project permitting and development/construction phase. Of course, these initiatives place a demand on government funds, but systems should be put in place to direct funding to them at times when the mining industry is buoyant by re-investing some of the increase in mining-derived revenues to further promote mining development. Development of common use infrastructure can then be viewed as an alternative, or better a complement, to the establishment of sovereign funds to ensure that fluctuating mining revenues are leveled put to a good use and not dissipated in extravagant politically motivated short-term initiatives at times of boom.

Alternatively above normal mining revenues at times of mineral booms may be directly contributed to an existing sovereign funds, provided its management has established provisions for the re-direction of specified proportions of its funds to be allocated specifically to the continued development of the minerals sector or of the regions affected by it in the country. An example of the latter "state-based" source of financial support was the 'royalties for regions' initiative of the Western Australian government, which hypothecated 25% of mineral royalty collections to carry out necessary capital works in the regions, other than urban areas, affected by mining operations.

While established common use infrastructure is a major incentive to invest, the state may still recover some of its investment by ways of user charges and in some cases of appropriate capital contributions from industry. An alternative is for government to join forces with private enterprise in establishing the required infrastructure through government-industry co-funding partnerships and build, own, operate and transfer (BOOT) mechanisms.

Governments must also use its authority to introduce or retain fiscal incentives for mining investors wisely in terms of their short and long term impacts on both itself and industry. Thus, while tax and royalty deferrals through accelerated depreciation and/or deferral of royalty payments may be desirable at the initial vulnerable stages of a project development, complete moratorium periods, including on CIT, royalties, import and export duty and etc. are undesirable from the point of view of government.

In addition, many governments have already or may consider introducing fiscal incentives that promote downstream processing in their respective countries. While this makes sense, government and industry must consider carefully the various risks involved, including:

- securing reliable electricity and water supplies. If these facilities are to be provided by government, than to ensure that public funds are not placed at risk, the mining industry users should pay appropriate headwork capital contributions related to the degree that they wish to secure the necessary capacity from government and enter into stringent take or pay type of contract for the supply services. Companies on the other hand will compare these conditions with the costs and risk involved in developing their own power and water supplies on site;
- correctly estimating and costing the necessary pre-emptive and reparatory measures to be put in place to ensure that the related environmental impacts can be contained within the required standards but with cognisance that these may be subject to rising societal expectations over time; and
- formulating stability agreements including clauses relating to security of tenure, appropriate labour laws ensuring adequate levels of labour security and other development conditions that would endure the test of time, since the establishment of processing facilities is both very expensive and long lived.

11.2.4 Simplify Their Tax Packages and Cut 'Red Tape' for Greater Clarity, Easier Administration and Lower Compliance Cost

Corporate income tax and mineral royalties (primarily weight or value based) represent the basic building blocks of the mining fiscal package of just about every jurisdiction. The CIT and royalty rates applied to the minerals industry, nonetheless, vary from jurisdiction to jurisdiction and from commodity to commodity and their proportional significance as a component of the government tax take also vary to achieve the different economic objectives of various jurisdictions. The weight/value based mineral royalties satisfy government's need for revenue adequacy and above all stability, while the CIT, being profit-based, tends to satisfy the objectives of economic efficiency and equity. In the end the choice of a fiscal regime is always a compromise between these two sets of largely mutually incompatible objectives, where pursuing revenue stability results in economically suboptimal outcomes in terms of higher cut-off grades, lower commercially feasible mining reserves and sterilisation of some of the mineral resources endowment. Notwithstanding, these two taxing instruments are generally well understood and accepted by industry and relatively easy for government to administer and industry to comply with.

However, in addition to these two basic components of the fiscal package, different jurisdictions have adopted a range of significantly more complex to administer profit or rent-based taxing mechanisms, government free carried equity participation and a plethora of 'nuisance' taxes and other imposts that can make mining tax administration and compliance unwieldly for both government and industry. A legitimate question is to what a degree these additional taxing instruments are essential, add value and/or could be dispensed of.

Starting with profit or rent-based mineral royalties, one may ask what they would achieve that is not achieved by the CIT. If the objective is to levy a higher level of rent from the mining sector relative to the rest of the economy, government, as in the case of Ghana, could achieve it by applying a higher rate of corporate income tax exclusively to this sector. In practice currently there are already very few jurisdictions using profit-base royalties and no resource rent taxes in operation in the mining sector.

The desirability of requiring mining companies to provide a free carried equity interest in their projects as a taxing mechanism is also debatable. This is because mining, being a chronically capital hungry industry, seldom pays dividends and when it does they tend to be meagre. In addition in general government cannot monetarise any capital gain by disposing of its mining shares on the market thus precluding this alternative source of funds. Attempts to circumvent these restrictions by for instance insisting on companies pay to government a set percentage of net cash flows in lieu of dividends are equivalent to and could be best accommodated through increases on the existing mineral royalties. Free carried equity participation is most frequently found in Africa and is probably a 'pride-of-ownership 'reaction to the pervasive foreign ownership of the continent's mineral resources during colonial times. Government equity participation also poses issues of conflict of interest for government officers appointed to the board of a company, who in theory should advance the interest of current shareholders even though these may not at times be aligned with that of government. The very process of appointing government representatives to a board directors is fraught with potential for ethical issues and often nepotism. Furthermore the practice may contribute to create or reinforce any perception that investors may have of incipient resources nationalism.

Administrative 'red tape' and duplication of effort are a major drain of resources from both government and industry. During the permitting, development and operating stages of a mining project industry has to deal with a wide range of different ministries and government departments, such as the Department of Mines, the Tax Office, the Department of the Environment etc., each of which has its unique regulatory policies, procedures and reporting requirements. These ministries/departments often operate in relative isolation and the boundaries of their respective responsibility and accountability is often blurred. Yet, with a few exceptions the performance conditions that attach to a mining title are pretty standard and appropriate protocols could be put in place so that industry could in the majority of cases deal with a single government institutions, the so-called 'one-stop shop', and only be referred to another specific department for a minority of individually unique and important issues. Government should fight this 'silos' mentality which can generally be traced back to individual ministries/departments protecting their power and political influence base, at its source and achieve greater coordination of the roles and responsibilities of the various ministries and department tasked with regulating the mining industry. It should ensure that relevant information provided by industry should be warehoused in a single reliable database, subject to scrupulous quality controls and accessible to all involved departments. It should also promote freely sharing the use of specialised skills and expertise between departments thus facilitating the resolution of more complex issues and expediting relevant processes. These actions should not detract from a continued effort to strengthen the specific fiscal administrative capacity, skills and systems to keep up with rapidly evolving industry's tax minimisation strategies.

11.2.5 Adhere to Current International Tax Reform Initiatives Designed to Prevent Base Erosion and Profit Shifting (BEPS)

As already discussed, the majority of tax issues in mining relate to the globalisation of the industry and its consequent restructuring that has resulted in erosion of the tax base of many mineral-rich nations and shifting of some of the profits generated by mining operations to foreign, generally low tax jurisdictions. However, while these issues are global in nature and require international tax reform to be addressed, implementation of these reforms needs to be enacted through amendments to the domestic tax legislation and review of the bilateral and multi-lateral international taxation agreements of the various countries which are signatories to the BEPS initiatives. To date over 100 international jurisdictions have agreed to implement the BEPS recommendations. The process of domestic tax reform, however, will require a significant effort on the side of many jurisdictions which may not currently have adequate resources or skills to implement it quickly and as a consequence the overall process may take a few years to be completed. To ensure a timely, orderly and unambiguous implementation of the BEPS recommendations it may be necessary for some jurisdictions to seek the assistance of international institutions and/or aid funds to engage specialised consultants.

11.2.6 Promote a More Harmonious Modus Operandi Viewing Mining Companies as Customers Not Adversaries

The tendency for the mining industry to engage specialised legal and tax experts to identify and exploit every opportunity and legal loophole for tax minimisation and their reluctance to provide promptly tax information when required by tax authorities has fostered mistrust and an adversarial attitude towards industry in many tax authorities. It has also resulted in many tax reform initiatives being formulated as knee-jerk reactions under an atmosphere of frustration with little or no consultation with industry. Springing one-sided decisions on industry without some degree of consultation often generates resistance, challenges and counter-productive results in general. A much more desirable approach is for government to adopt a general policy of 'no surprises', that is to say that no new mining fiscal initiative should be introduced without having given industry sufficient information to understand and comment about and ideally provide constructive criticism to its formulation. Government should identify, inform and consult with all the stakeholders potentially affected by a proposed reform and consider their comments obtained both through their representative bodies and individually. This does not mean that government should capitulate to industry's objections or accede to their demands without clear consideration of their logic or merit, but that industry should clearly understand the rationale for the changes and be gradually led to accept them and not taken by surprise. Taxpayers, if properly handled, can still make a valuable contribution to policy formulation even if they do not like the general thrust and potential consequences of a fiscal initiative. One of the authors had a most favourable experience of this approach when involved through the taxation committee of an industry representative body in progressively reviewing and commenting on the 'exposure draft' of the proposed MRRT prior to its introduction in Australia. It is fair to say that very few, if any, participants in the mining industry in Australia other than the major four companies, liked the proposed MRRT, but the industry did nonetheless contribute to making the related legislation and regulations practically enforceable by analysing their administrative impact and practicality from the point of view of industry compliance.

While there will always be a few rogue individuals who will try to game the system, tax authorities will be better served by assuming that the majority, while minimising their taxes, will try to operate within the bounds of the law and that tax authorities are in the business of providing a service not just to government but also to individual taxpayers. A change in philosophy from viewing taxpayers as adversaries to 'customers' goes hand in hand with systemic changes that will gradually prevent rather than punish non-compliance and tax avoidance.

11.2.7 Ensure Ready Access to and Availability of Key Officials to Provide Mining Companies with Relevant Advice and Prompt Decisions

Exploration and mining companies face many challenges, particularly so when they enter for the first time into an unfamiliar country to commence exploration or extractive activities. The challenges being faced are often exacerbated when government officials do not make themselves available to provide country guidance and timely responses to requests for information by a new entrant in their market. It is important for governments to remember that the quantum of expenditure that may ultimately be incurred in their country in the commercial development of a mineral property is significant and therefore warrants not only the attention and interest of the local authorities but also their support.

Countries that have a good reputation through previous supportive industry engagements will find that mining companies have a predisposition to investing in them and also tend to develop the depth of market to support attracting equity capital, while the opposite is true. Countries new to mining investment have an early opportunity to ensure they present themselves and develop a perception of being investor-friendly.

It is important to keep in mind that alternative investment jurisdictions exist and that companies will weigh up any investment opportunity against the ease or difficulty of doing business in various countries compared with the alternative investment climate, returns and challenges in other countries. If government officials are unapproachable, continuously unavailable or demonstrate a general disinterest, the investment opportunities that could be pursued in their country may be overlooked.

11.3 What the Mining Industry Could Do

11.3.1 Carefully Analyse the Strength and Weaknesses of the Fiscal Regimes of Various Possible Investment Destinations Before Taking the Plunge

As already pointed out, geological prospectivity is the main factor to invest in a country with the attractiveness of its mining taxation regime a close second. So why a company may be forgiven for embarking into exploration prior to a thorough analysis of a country's fiscal regime particularly if it attempts to ripe the benefits of 'first mover', there is no justification to embark in the major capital investment necessary to develop a mine without a clear understanding of the fiscal consequences of such an action. Such an analysis should ideally be complemented with consultations with the local tax authorities for clarification and exploration of the terms of a possible stabilisation agreements. These individual companies' assessments would contribute to the growing body of knowledge about the attractiveness of the mining fiscal regime of various countries and in turn provide a feedback loop to their government as to how to position their countries in terms of investment attraction and government-industry interaction.

11.3.2 More Effectively Communicate with Government Both Collectively and at the Individual Company Level to Inform and Influence Mining Fiscal Policy Formulation and Administration

Communication with the tax authorities should not be limited primarily to when specific issues arise, but should be a continuous process of mutual exchange of information and support in the process of policy formulation and framing of administrative procedures. This process of communication and cooperation should take place at both the collective industry level through representative bodies and at the individual taxpayer level. As for many other industries, the mining industry tends to be composed of a handful of major companies and a larger number of less influential small and medium enterprises (SMEs) and individual prospectors. Major companies are generally better known by and have a relatively high level of clout in government because they are generally handled by dedicated 'large taxpayers' sections in the tax office. They are also the main contributors and most influential members of formal industry representative bodies, such as the Chamber of Mines that cater mainly for their interests. These may not in all cases coincide with the SMEs' interests which may require the constitution of a separate representative/lobbying body. It is highly desirable that industry representative bodies should include a standing 'Mining tax committee', composed of industry practitioners and tax consultants, to monitor and inform members about developments in the fiscal area and to be a vehicle for communication with tax authorities. By not serving the exclusive interests of an individual taxpayer the tax committee and the tax authorities are in a position to approach issues in a more neutral and analytical manner and in doing so to engender a degree of mutual openness and trust. Ideally industry should create a degree of dependence on the side of the tax authority for reliable information in support of policy formulation relating to mining.

11.3.3 Engage as Far as Possible the Support of Local Communities Affected by the Project Through Local Employment and Procurement and Other Visible Socio-Economic Initiatives

Mineral occurrences are discovered at or close to surface, at depth or even under water. Owing to the population expansion, an ever increasing number of mineral opportunities may have a direct or indirect impact on local and regional communities that are resident within the mineralised areas, where conflicting land-uses may arise. As a consequence, mining companies must effectively and sensitively engage with any party that may potentially be impacted early on in the exploration phase in order to obtain the necessary local buy-in and support for the future exploration and development of a possible mineral discovery.

The buy-in and support may require addressing some of the future requirements and expectations of the affected parties. The strategy must be to focus primarily on initiatives that would align as far as possible the interests of the affected communities with those of the company and may manifest themselves as:

- the local community being granted an equity position in the operation;
- member s of the local population being employed to provide assistance during the exploration phase of the programme, receiving payment for typically unskilled services including site-clearing, drilling assistance, navigation, etc.;
- educational support including provision of schools for local children, and of adult based education training (ABET) programmes, improving the literacy and numeracy of the local population, as well as trade-based training and apprenticeships, and bursary schemes for local residents to attend institutions of higher learning and entrepreneurial training employment of some of the above trainees in the operations and/or supporting their establishing local organisations for the provision of goods and services to the mining operations.

The various costs involved in creating the above opportunities for local community residents are deemed to be operating costs of the project and therefore are tax-deductible.

To the extent that, during the exploration phase, most of the funds being used for are equity-sourced and expensive and that the exploration company is not generating any revenue, the range of activities that may be funded during exploration will be limited compared to those that will become possible once the decision to develop a possible discovery is made.

11.3.4 Improve the Realism of the Feasibility Studies and Promptly Inform Government of Any Significant Departure from the Initial Plan During Development and Subsequent Operations

A major area of contention between government and the mining industry has been the degree of variance between the forecasts underpinning feasibility studies and the actual unfolding of events during mining operations. This is particularly the case when mineral royalties and CIT collections, local labour requirements and procurements do not live up to the expectations of government which were largely influenced by the project feasibility study submitted while negotiating the granting of the project mining contract. While government generally understands that all the feasibility study inputs are subject to a degree of uncertainty and tolerates the fact that, within limits, some of the actual outcomes may be a variance with plans and budgets, there have been many cases where project developments deliberately progressed along lines and produced results vastly different from those assumed in their feasibility studies. This, combined with the effect of the inherent volatility of mineral commodity prices, has created significant resentment in some government circles and pressure to formulate policies that would introduce minimum royalties and CIT collections floor in line with the amount that would have been collected if the forecasts in the relevant feasibility studies had actually eventuated. While this approach is neither a fair or practical solution to the problem and should not be adopted, something needs to be done to ensure that problems in this area are addressed as they emerge and not be allowed to fester. Industry could deflate the problem by defining and agreeing with government as to what constitute acceptable probabilistic departures from the outcomes as planned in the feasibility study, but promptly alert government of any impending more significant departure from plans, particularly if these are attributable to a deliberate change of plans or scope of the project. The latter course of action, which is often prompted by an unwarranted interpretation of certain fiscal incentives such as the tax holidays described in Chap. 5, may bring about the need for a renegotiation of some of the terms of the mining contract.

11.3.5 Establish a More Transparent and Cooperative Relationship with the Local Tax Authority by Being Upfront with Information Relevant to Potential Inquiries

Much of the impunity of the tax avoidance practices of some of the mining MNEs in the past was attributable to the inability of the tax authorities in jurisdictions hosting the mining operations to obtain sales revenue and cost information for auditing of transactions occurring in foreign jurisdictions. Even if information were theoretically obtainable by virtue of an exchange of information condition in the double taxation agreement with a relevant foreign jurisdiction, the process tended to be long-protracted and the inevitable delays mitigated against positive resolution of fiscal disputes. However, in light of the dramatically increasing rapidity of digital information exchange between tax jurisdictions under the multilateral instruments and more stringent documentation and reporting requirements brought about by BEPS, it will become very difficult for taxpayers to withhold, obfuscate or delay the release information to frustrate audits. Mining companies will need to resign themselves to the inevitability of having in the future to pay a more appropriate level of tax. Under this new set of circumstances it will pay for individual mining companies to improve their relationship with tax authorities by adopting a more open and transparent approach in their dealing with them. This would include providing the tax authorities with relevant briefings about new initiatives, seeking their advice and rulings in matters of tax planning and of proposed corporate restructuring and a high level of promptness in submitting any documentation required by the tax office ensuring its clarity and relevance.

11.3.6 Accept that the BEPS Action Plans Have Now Been Accepted by a Majority of Jurisdictions and that Many Tax Benefits Derived from Artificial Global Structures and Liberal Interpretations of Transfer Prices Rules Are Rapidly Closing

It is fair to assume that prior to the broad international acceptance and progressive implementation of the BEPS Actions recommendations the predominant focus of MNEs restructuring had been on tax minimisation and that, as a consequence, the potential productivity benefits and operational synergies of alternative structures had in many cases a lower or no influence on the final shape of mining MNEs. Since 2015, however, the international fiscal landscape has drastically changed with the majority of the tax minimisation opportunities and other loopholes generated by the globalisation of the mining industry being reduced or completely eliminated. Under this changed set of circumstances, the mining industry will progressively recognise the inevitability of having to pay collectively more taxes at the MNE consolidated level. To offset this increase in costs the industry will have to explore other potential sources of savings and new avenues for productivity improvements. This may include MNEs once again reviewing their global structures with a view to achieve logistical and other operational synergies rather than being primarily driven by tax considerations. Consequently MNEs may consider establishing their marketing hubs in locations that offer potential benefits in terms of being closer to markets and customers or of providing superior harbours, storage and shipping facilities. Similarly they may attempt to achieve effective critical mass and operational efficiency by consolidating most of their non-routine and high value-adding specialized engineering, scientific, technical and R&D capabilities, as well as related intangibles assets, into one or a few subsidiaries located in countries where there is good research infrastructure, and where it is easier to attract and retain the required specialized technical skills. In this next phase of corporate restructuring MNEs will also need to keep in mind that the increasing reliance on digital technology will have a significant and continuing impact on the nature of service delivery and on the optimal composition of its skill inventory and less on the choice of location for its offices and specialised staff.

11.3.7 Behave in an Environmentally and Socially Responsible Manner

While this may sound intuitively obvious, it is important that explorers and mine operators endeavour to behave responsibly in terms of both the environment and from a social perspective. This is not only important from the specific operator's standpoint, but also in terms of contributing to improve and maintain the wider reputation and good standing of the global mining industry at large as being environmentally responsible. It only takes one isolated incident to bring down the reputation of the whole sector, as in the case of the tailings dams failures that occurred in Brazil over the course of 2019.

Companies must be seen to be actively looking after the environment and must progressively rehabilitate their operational disturbance whenever possible. Environmental degradation and pollution prevention measures must be implemented continuously to ensure that potential permanent scarring is avoided or at least minimised. This must relate to environmental impacts affecting land, water, subterranean features and the atmosphere.

Social responsibility is another important area that companies must manage carefully. The various stakeholders, including local and regional communities, need to be encouraged to support the operating activities and this may be best achieved through their involvement. Involvement in this context includes the factors outlined in 7.3.3 above and extends further into more-difficult-to-measure types of involvement culminating in mutual trust and respect between parties and a general positive relationship held between the company and the community.

11.4 Conclusive Remarks

In summary, government has a vast range of choice when it comes to the taxing instrument that they could adopt to shape the fiscal package to apply to the mining industry. All jurisdictions have adopted mineral royalties, primarily specific and *ad valorem*, and corporate income tax (CIT) as the basic blocks of their fiscal packages, but differ on the degree of emphasis they place on either instrument. Emphasis on mineral royalties provides greater revenue stability and ease of administration at the cost of economic efficiency and equity and *vice versa* for CIT. In addition, capital gain tax, VAT/GST, custom and excise duties, withholding tax and a plethora of less significant 'nuisance' taxes are applicable to the majority of jurisdictions, but to varying individual degrees. Finally many jurisdictions, particularly in developing countries insist on being granted free carried equity participation in mining projects as well as demanding a range of quasi-taxes in the form of commitments to fund a range of socio-economic initiatives. This multiplicity of imposts complicates the tax administration and increases the compliance cost for both government and industry. It also makes it rather difficult for potential mining investors to compare the total tax take of different investment destinations and makes direct comparison of the royalty and CIT rates of different jurisdictions practically meaningless.

The book concludes that there is a real need to simplify some of the fiscal regimes applying to mining by reconsidering the need for government equity participation, eliminating some of the nuisance taxes and to cut 'red tape' by establishing sound administrative protocols to smooth the interaction among the range of ministries and departments involved in regulating the industry, ideally making it possible to address the majority of the issues affecting the industry through a 'one-shop' system.

Another area of industry concern when considering investing in a jurisdiction is the degree to which it can rely on the political and administrative system remaining stable over the life of their project/investment. This concern is generally addressed by negotiating a 'stability agreement' which guarantees security of title and no future changes in the fiscal conditions applying to a project. Past experience with stability agreement has not always been positive, primarily for both government's and industry's inability to think realistically decades in the future, the high degree of uncertainty surrounding the mining industry and the rigidity of the agreements that may be only changed by mutual consent of the two parties and, sometimes, require parliamentary ratification.

The book suggests that the terms of future stabilisation agreements should be reviewed to allow their renegotiation at regular albeit reasonably lengthy (10–15 years) intervals, or by introducing 'triggers' for renegotiation when certain pre-set parameters, e.g. commodity prices, are exceeded.

Investment decisions need to be informed by better financial modelling of fiscal arrangements with the results shared, as widely as possible while respecting any confidentiality requirements imposed by regulatory bodies such as stock exchanges, securities regulators, etc. Of particular importance is the need to ensure that relevant government officials fully understand the financial and economic implications of their policies and actions in terms of the project expected outcomes, flow and distribution of benefits.

Increasing effort is being placed through progressive acceptance of the BEPS recommendations on reducing the degree to which the tax bases of mineral-rich developing countries hosting mining operations has recently been eroded with profits shifted overseas as an effect of globalisation with profits. This has given rise to a rapidly changing mining taxation landscape in most mining jurisdictions. While it is generally accepted that tightening up the control measures to avoid BEPS occurring is justified, it must be borne in mind that this will ultimately increase the tax payable by MNEs at the consolidated level and that this in combination with continuous and in some cases unwarranted fiscal policy changes may actually affect the investment attractiveness of the mining industry in general and cause some of the FDI to be redirected from it and potentially dry up.

Increased communication and transparency is also needed between government and industry on regulatory and fiscal matters. This should be a two way street with industry being more open and prompt in the provision of relevant information relating to tax authorities inquiries and audits and government keeping industry continuously informed and involved in the formulation of new policies and regulations or the amendment of existing ones affecting it avoiding as far as possible taking industry by surprise. Although this 'no surprise' approach does not imply government having to accede to any industry objection or demand, it will nonetheless foster greater cooperation and mutual trust between the parties and ultimately sounder and more practically implementable decisions.

Take Away Points

Government's point of view

- Governments are mandated, in part, to secure direct financial returns, derived through mechanisms including taxes, mineral royalties, import/export duties, tariffs, dividends on carried or free carried equity interests in mining projects, and other accrual mechanisms,
- Governments also encourage or legislate indirect monetary and non-monetary benefits, derived through mechanisms including local and broader employment multipliers, subsidised local equity participation, skills transfer programmes, education programmes, provision of specific infrastructure, entrepreneurial programmes, establishment of secondary industry, diversifying services offered by main-stream suppliers and other economically beneficial activities.
- In support of developing the mining sector, Governments promote mineral exploration, communicate with industry, provide appropriate mineral policies, ensure tax packages are clear and transparent, adhere to international standards designed to prevent base erosion and profit shifting, work amicably with industry and are timeous in their advice and responses.

Mining industry's point of view

- Since mining companies can choose where (jurisdiction) to invest, they assess and analyse fiscal regimes for attractiveness, typically communicate with Governments around their intentions, engage with local communities and assist in generating socio-economic initiatives, provide realistic plans and potential outcomes relating to their intended activities, collaborate with the tax authorities, adhere to BEPS initiatives and conduct themselves in an environmentally and socially responsible manner.

Appendices

Appendix A: Guiding Principles for Durable Extractive Contracts

Following extensive stakeholders' consultation the OECD Development Centre has in 2019 published the above Guiding Principles designed "to assist host governments and investors using them to (1) structure their on-going relationship in an integrated manner to promote long term sustainable development, while attracting and sustaining investment; (2) foster alignment of expectations and convergence towards agreed objectives; (3) provide mechanisms that can accommodate and respond in a predictable manner to potentially significant changes in circumstances; (4) build trust to strengthen mutual confidence and reduce risk for both parties; (5) ensure a fair share for all parties to the contract and optimise the value from resource development through equitable, sustainable and mutually beneficial contracts and operations". The following eight principles, while not officially representing the views of the OECD were nonetheless based on wide consultations with its member and non-member countries, the mining industry and various other institutions and interested parties. The principles are not presented in hierarchical order, interact with each other and should therefore be considered together.

1. Durable extractive contracts are aligned with the long-term vision and strategy, defined by the host government on how the extractive sector can fit into and contribute to broader sustainable development objectives.
2. Durable extractive contracts are anchored in a transparent, constructive long-term commercial relationship and operational partnership between host governments, investors and communities, to fulfil agreed and understood objectives based on shared and realistic expectations that are managed throughout the life-cycle of the project.
3. Durable extractive contracts balance the legitimate interests of host governments, investors, and communities, with due account taken, where relevant, of the

E. Lilford, P. Guj, *Mining Taxation*, Modern Approaches in Solid Earth Sciences 18,
https://doi.org/10.1007/978-3-030-49821-4

specific rights of affected indigenous peoples recognised under applicable international and/or national law.

4. Durable extractive contracts seek to optimise the value from resource development for all stakeholders, including economic, social and environmental outcomes.

   To the extent not covered by applicable international and/or national law, durable extractive contracts provide for the identification and management of potential adverse environmental, health, safety and social impacts of the extractive project and establish clear roles and responsibilities for the host government and the investor for the prevention, mitigation and remediation of those impacts, in consultation with affected communities.
5. Durable extractive contracts are negotiated and based on the continued sharing, in good faith, of key financial and technical data to build a common understanding of the performance, main risks and opportunities of the project throughout its lifecycle.
6. Durable extractive contracts operate in a sound investment and business climate and should be underpinned by a fair, transparent and clear legal and regulatory framework and enforced in a non-discriminatory manner.
7. Durable extractive contracts are consistent with applicable laws, applicable international and regional treaties, and anticipate that host governments may introduce bona fide, non-arbitrary, and non-discriminatory changes in law and applicable regulations, covering non-fiscal regulatory areas to pursue legitimate public interest objectives. The costs attributable to compliance with such changes in law and regulations, and wholly, necessarily and exclusively related to project specific operations, should be treated as any other project costs for purposes of tax deductibility, and cost recovery in production sharing contracts. If such changes in law and/or applicable regulations result in the investor's inability to perform its material obligations under the contract or if they lead to a material adverse change that undermines the economic viability of the project, durable extractive contracts require the parties to engage in good faith discussions which might eventually lead the parties to agree to renegotiate the terms of the contract.
8. Durable extractive contracts are underpinned by a fiscal system that is consistent with the governments' overall economic and fiscal objectives and provides a fair sharing of financial benefits between the investor and the host government, taking into consideration the potential risks, rewards, and country circumstances. As there is no one ideal fiscal regime, each host government needs to identify the optimal mix of fiscal instruments and terms to meet its objectives.

   A predictable fiscal regime that includes responsive terms defined in legislation and/or the contract to adjust the allocation of overall financial benefits between host governments and investors to variables that affect project profitability (such as variance in commodity prices, costs, production volume, or resource quality) contributes to the long-term sustainability of extractive contracts and reduces the incentives for either party to seek renegotiation of terms. Host governments need to generate financial benefits from the extraction of their resources. Durable extractive contracts avoid sustained periods of commercial production with little or no revenue flows to the government.

Appendix B: Country Risk Classifications of the Participants to the Arrangement on Officially Supported Export Credits (Source OECD: http://www.oecd.org/trade/topics/export-credits/documents/cre-crc-current-english.pdf)

Valid as of: **18 October 2019**

	Country Code ISO Alpha 3	Country Name [1]	Classification		Notes
			Previous	Current Prevailing	
1	AFG	Afghanistan	7	7	
2	ALB	Albania	5	5	
3	DZA	Algeria	4	4	
4	AND	Andorra	-	-	(9)
5	AGO	Angola	6	6	
6	ATG	Antigua and Barbuda	7	7	(8)
7	ARG	Argentina	6	7	
8	ARM	Armenia	6	6	
9	ABW	Aruba	4	4	
10	AUS	Australia	-	-	(6)
11	AUT	Austria	-	-	(6) (7)
12	AZE	Azerbaijan	5	5	
13	BHS	Bahamas	3	3	
14	BHR	Bahrain	5	5	
15	BGD	Bangladesh	5	5	
16	BRB	Barbados	-	-	(5)
17	BLR	Belarus	6	6	
18	BEL	Belgium	-	-	(6) (7)
19	BLZ	Belize	-	-	(5)
20	BEN	Benin	6	6	(8)
21	BTN	Bhutan	6	6	
22	BOL	Bolivia	5	5	
23	BIH	Bosnia and Herzegovina	7	7	
24	BWA	Botswana	2	2	
25	BRA	Brazil	5	5	
26	BRN	Brunei Darussalam	-	-	(5)
27	BGR	Bulgaria	3	3	
28	BFA	Burkina Faso	7	7	(8)
29	BDI	Burundi	7	7	
30	CPV	Cabo Verde	6	6	
31	KHM	Cambodia	6	6	
32	CMR	Cameroon	6	6	(8)
33	CAN	Canada	-	-	(6)
34	CAF	Central African Republic	7	7	(8)
35	TCD	Chad	7	7	(8)
36	CHL	Chile	-	-	(6)
37	CHN	China (People's Republic of)	2	2	
38	COL	Colombia	4	4	
39	COM	Comoros	-	-	(5)
40	COG	Congo	7	7	(8)
41	CRI	Costa Rica	3	3	
42	CIV	Côte d'Ivoire	6	6	(8)
43	HRV	Croatia	4	4	
44	CUB	Cuba	7	7	
45	CUW	Curaçao	5	5	
46	CYP	Cyprus	-	-	(2) (7)
47	CZE	Czech Republic	-	-	(6)
48	PRK	Democratic People's Republic of Korea	7	7	

	Country Code ISO Alpha 3	Country Name [1]	Classification Previous	Classification Current Prevailing	Notes
49	COD	Democratic Republic of the Congo	7	7	
50	DNK	Denmark	-	-	(6)
51	DJI	Djibouti	7	7	
52	DMA	Dominica	-	-	(5)
53	DOM	Dominican Republic	4	4	
54	ECU	Ecuador	6	6	
55	EGY	Egypt	5	5	
56	SLV	El Salvador	5	5	(8)
57	GNQ	Equatorial Guinea	7	7	(8)
58	ERI	Eritrea	7	7	
59	EST	Estonia	-	-	(6) (7)
60	SWZ	Eswatini	6	6	
61	ETH	Ethiopia	7	7	
62	FJI	Fiji	5	5	
63	FIN	Finland	-	-	(6) (7)
64	FRA	France	-	-	(6) (7)
65	GAB	Gabon	6	6	(8)
66	GMB	Gambia	7	7	
67	GEO	Georgia	6	6	
68	DEU	Germany	-	-	(6) (7)
69	GHA	Ghana	6	6	
70	GRC	Greece	-	-	(6) (7)
71	GRD	Grenada	-	-	(5)
72	GTM	Guatemala	4	4	
73	GIN	Guinea	7	7	
74	GNB	Guinea-Bissau	7	7	(8)
75	GUY	Guyana	6	6	
76	HTI	Haiti	7	7	
77	HND	Honduras	5	5	
78	HKG	Hong Kong (China)	2	2	
79	HUN	Hungary	-	-	(6)
80	ISL	Iceland	-	-	(6)
81	IND	India	3	3	
82	IDN	Indonesia	3	3	
83	IRN	Iran	6	6	
84	IRQ	Iraq	7	7	
85	IRL	Ireland	-	-	(6) (7)
86	ISR	Israel	-	-	(3) (6)
87	ITA	Italy	-	-	(6) (7)
88	JAM	Jamaica	6	6	
89	JPN	Japan	-	-	(6)
90	JOR	Jordan	5	5	
91	KAZ	Kazakhstan	5	5	

	Country Code ISO Alpha 3	Country Name [1]	Classification		Notes
			Previous	Current Prevailing	
92	KEN	Kenya	6	6	
93	KIR	Kiribati	-	-	(5)
94	KOR	Korea	-	-	(6)
95		Kosovo	7	7	(4)
96	KWT	Kuwait	2	2	
97	KGZ	Kyrgyzstan	7	7	
98	LAO	Lao People's Democratic Republic	7	7	
99	LVA	Latvia	-	-	(6) (7)
100	LBN	Lebanon	7	7	
101	LSO	Lesotho	6	6	
102	LBR	Liberia	7	7	
103	LBY	Libya	7	7	
104	LIE	Liechtenstein	-	-	(5)
105	LTU	Lithuania	-	-	(7)
106	LUX	Luxembourg	-	-	(6) (7)
107	MAC	Macau (China)	2	2	
108	MDG	Madagascar	7	7	
109	MWI	Malawi	7	7	
110	MYS	Malaysia	2	2	
111	MDV	Maldives	6	6	
112	MLI	Mali	7	7	(8)
113	MLT	Malta	-	-	(7)
114	MHL	Marshall Islands	-	-	(5)
115	MRT	Mauritania	7	7	
116	MUS	Mauritius	3	3	
117	MEX	Mexico	3	3	
118	FSM	Micronesia	-	-	(5)
119	MDA	Moldova	7	7	
120	MCO	Monaco	-	-	(9)
121	MNG	Mongolia	6	6	
122	MNE	Montenegro	7	7	
123	MAR	Morocco	3	3	
124	MOZ	Mozambique	7	7	
125	MMR	Myanmar	6	6	
126	NAM	Namibia	5	5	
127	NRU	Nauru	-	-	(5)
128	NPL	Nepal	6	6	
129	NLD	Netherlands	-	-	(6) (7)
130	NZL	New Zealand	-	-	(6)
131	NIC	Nicaragua	7	7	
132	NER	Niger	7	7	(8)
133	NGA	Nigeria	6	6	
134	MKD	Northern Macedonia	5	5	
135	NOR	Norway	-	-	(6)
136	OMN	Oman	4	4	

	Country Code ISO Alpha 3	Country Name [1]	Classification		Notes
			Previous	Current Prevailing	
137	PAK	Pakistan	7	7	
138	PLW	Palau	-	-	(5)
139	PAN	Panama	3	4	(8)
140	PNG	Papua New Guinea	6	6	
141	PRY	Paraguay	5	5	
142	PER	Peru	3	3	
143	PHL	Philippines	3	3	
144	POL	Poland	-	-	(6)
145	PRT	Portugal	-	-	(6) (7)
146	QAT	Qatar	3	3	
147	ROU	Romania	3	3	
148	RUS	Russia	4	4	
149	RWA	Rwanda	6	6	
150	KNA	Saint Kitts and Nevis	-	-	(5)
151	LCA	Saint Lucia	-	-	(5)
152	VCT	Saint Vincent and the Grenadines	-	-	(5)
153	WSM	Samoa	-	-	(5)
154	SMR	San Marino	-	-	(9)
155	STP	Sao Tome and Principe	-	-	(5)
156	SAU	Saudi Arabia	2	2	
157	SEN	Senegal	5	5	(8)
158	SRB	Serbia	5	5	
159	SYC	Seychelles	-	-	(5)
160	SLE	Sierra Leone	7	7	
161	SGP	Singapore	0	0	
162	SXM	Sint Maarten	-	-	(5)
163	SVK	Slovak Republic	-	-	(6) (7)
164	SVN	Slovenia	-	-	(6) (7)
165	SLB	Solomon Islands	-	-	(5)
166	SOM	Somalia	7	7	
167	ZAF	South Africa	4	4	
168	SSD	South Sudan	7	7	
169	ESP	Spain	-	-	(6) (7)
170	LKA	Sri Lanka	6	6	
171	SDN	Sudan	7	7	
172	SUR	Suriname	6	6	
173	SWE	Sweden	-	-	(6)
174	CHE	Switzerland	-	-	(6)
175	SYR	Syrian Arab Republic	7	7	
176	TWN	Chinese Taipei	1	1	
177	TJK	Tajikistan	7	7	
178	TZA	Tanzania	6	6	
179	THA	Thailand	3	3	
180	TLS	Timor-Leste	6	6	
181	TGO	Togo	6	6	(8)

	Country Code ISO Alpha 3	Country Name [1]	Classification		Notes
			Previous	Current Prevailing	
182	TON	Tonga	-	-	(5)
183	TTO	Trinidad and Tobago	3	3	
184	TUN	Tunisia	5	5	
185	TUR	Turkey	5	5	
186	TKM	Turkmenistan	6	6	
187	TUV	Tuvalu	-	-	(5)
188	UGA	Uganda	6	6	
189	UKR	Ukraine	7	7	
190	ARE	United Arab Emirates	2	2	
191	GBR	United Kingdom	-	-	(6)
192	USA	United States	-	-	(6)
193	URY	Uruguay	3	3	
194	UZB	Uzbekistan	5	5	
195	VUT	Vanuatu	-	-	(5)
196	VEN	Venezuela	7	7	
197	VNM	Viet Nam	4	4	
198	PSE	West Bank and Gaza	7	7	
199	YEM	Yemen	7	7	
200	ZMB	Zambia	7	7	
201	ZWE	Zimbabwe	7	7	

Notes

(1) *This document and any map included herein are without prejudice to the status of or sovereignty over any territory, to the delimitation of international frontiers and boundaries and to the name of any territory, city or area.*

(2) *Footnote by Turkey*
The information in this document with reference to "Cyprus" relates to the southern part of the Island. There is no single authority representing both Turkish and Greek Cypriot people on the Island. Turkey recognises the Turkish Republic of Northern Cyprus (TRNC). Until a lasting and equitable solution is found within the context of the United Nations, Turkey shall preserve its position concerning the "Cyprus issue".
Footnote by all the European Union Member States of the OECD and the European Commission
The Republic of Cyprus is recognised by all members of the United Nations with the exception of Turkey. The information in this document relates to the area under the effective control of the Government of the Republic of Cyprus.

(3) *The statistical data for Israel are supplied by and under the responsibility of the relevant Israeli authorities. The use of such data by the OECD is without prejudice to the status of the Golan Heights, East Jerusalem and Israeli settlements in the West Bank under the terms of international law.*

(4) *This designation is without prejudice to positions on status, and is in line with United Nations Security Council Resolution 1244/99 and the Advisory Opinion of the International Court of Justice on Kosovo's declaration of independence.*

(5) *Currently not reviewed or classified.*

(6) *High Income OECD Country not reviewed or classified.*

(7) *High Income Euro Area Country not reviewed or classified.*

(8) *According to article 26 of the Arrangement, for all countries classified through the county risk classification methodology, the risk of the sovereign is also assessed in order to identify, on an exceptional basis, sovereigns that are: (1) not the lowest risk obligor in the country and (2) whose credit risk is significantly higher than country risk.*

(9) *European micro-state that uses the euro through monetary agreement with the European Union not reviewed or classified.*

Appendix C: Extract from ACIL Allen Consulting (2015) Economic Evaluation of the Exploration Incentive Scheme Offered by the Geological Survey of Western Australia

The report key findings are summarised below:

- there are externality issues associated with mineral exploration that result in under investment in grass root exploration. As such, well designed government programs that address these externality issues—such as the EIS—can increase community welfare
- for the sample period there was a strong private sector exploration response to the EIS and the additional exploration activity generated net benefits for Western Australia
- the expected combined Western Australian private sector and government sector benefit, in net present value terms, is \$23.7M for every \$1M invested in the EIS. This finding is consistent with the results reported from other studies
- per \$1M invested in the EIS, the expected benefits are:
  - the direct financial impact of the additional exploration activity stimulated by the EIS of \$10.3M
  - the employment impact of the additional exploration activity stimulated by the EIS, which is 12.5 FTE positions in minerals exploration for 3 years
  - the financial impact of the mine site construction phase, which is estimated to be \$3.9M
  - the employment impact of the mine site construction phase, which is estimated to result in 27.6 FTE positions for the 2 year construction phase
  - the share of net private sector wealth that accrues to Western Australians following the successful development of a mine, which is estimated to be \$3.3M
  - the employment created during the production phase which is 7.4 FTE for 13 years
  - the net present value of additional government revenue generated by royalty and pay-roll tax payments of \$6.2M
- for low price scenarios the expected value of an average drilling campaign is negative. As such, the expected private sector exploration response to incentives under a low price scenario is expected to be modest; as are the benefits that accrue to the state
- for higher price scenarios, market conditions make it possible to fund the development of very large mining projects. These projects operate for long periods, potentially delivering royalty, employment, and net wealth creation benefits to Western Australia for decades. Under the high price scenario the expected net present value of every \$1M invested in the EIS is \$38.3M.

Appendix D: Cash Flow Model Summary

Year		0	1	2	3	4	5	6	7	8	9	10	11	12	13	14	15
Geological resource (Mt)	32.00	32.00	32.00	32.00	32.00	32.00	32.00	32.00	32.00	32.00	32.00	32.00	32.00	32.00	32.00	32.00	32.00
Geological grade (Au g/t)	4.50	4.50	4.50	4.50	4.50	4.50	4.50	4.50	4.50	4.50	4.50	4.50	4.50	4.50	4.50	4.50	4.50
Conversion factor resources to reserves (%)	86%	86%	86%	86%	86%	86%	86%	86%	86%	86%	86%	86%	86%	86%	86%	86%	86%
Operational mining losses (%)	1.5%	1.5%	1.5%	1.5%	1.5%	1.5%	1.5%	1.5%	1.5%	1.5%	1.5%	1.5%	1.5%	1.5%	1.5%	1.5%	1.5%
Dilution by weight	11.0%	11.0%	11.0%	11.0%	11.0%	11.0%	11.0%	11.0%	11.0%	11.0%	11.0%	11.0%	11.0%	11.0%	11.0%	11.0%	11.0%
Grade of dilutant (Au g/t)	1.00	1.00	1.00	1.00	1.00	1.00	1.00	1.00	1.00	1.00	1.00	1.00	1.00	1.00	1.00	1.00	1.00
Diluted Mining Reserves (Mt)	30.09	30.09	30.09	30.09	30.09	30.09	30.09	30.09	30.09	30.09	30.09	30.09	30.09	30.09	30.09	30.09	30.09
Diluted grade g/t Au	4.15	4.15	4.15	4.15	4.15	4.15	4.15	4.15	4.15	4.15	4.15	4.15	4.15	4.15	4.15	4.15	4.15
Annual ore production (Mt) (as reported)					0.10	2.00	3.00	3.00	3.00	3.00	3.00	3.00	3.00	3.00	3.00	0.99	0.00
Recovery (%)	91.0%																
Gold production Moz					0.012	0.243	0.365	0.365	0.365	0.365	0.365	0.365	0.365	0.365	0.365	0.120	
Price US$/oz Au	1,320				1,432.5	1,472.2	1,512.9	1,554.7	1,597.7	1,641.9	1,687.3	1,733.9	1,781.9	1,831.1	1,881.7	1,933.8	
Price A$/oz Au	1,750				1,899.2	1,951.7	2,005.7	2,061.1	2,118.1	2,176.7	2,236.9	2,298.7	2,362.3	2,427.6	2,494.7	2,563.7	
Sales of gold A$M					23.08	474.30	731.13	751.34	772.12	793.47	815.41	837.95	861.12	884.93	909.40	308.09	
REVENUE $M					23.08	474.30	731.13	751.34	772.12	793.47	815.41	837.95	861.12	884.93	909.40	308.09	85.35
Royalty as % of gross revenue $M @	**7.0%**				−1.62	−33.20	−51.18	−52.59	−54.05	−55.54	−57.08	−58.66	−60.28	−61.95	−63.66	−21.57	
Operating Expenses inputs in $(Year 0) then escalated																	
Fixed costs $M/p.a.	10.00				−10.77	−11.04	−11.32	−11.60	−11.89	−12.19	−12.50	−12.81	−13.13	−13.46	−13.80	−14.15	
Mining: Ore $M @ ($/t)	34.00				−3.66	−75.09	−115.46	−118.36	−121.33	−124.37	−127.50	−130.70	−133.98	−137.34	−140.79	−47.58	
Labour	38.00				−4.09	−83.92	−129.04	−132.28	−135.60	−139.01	−142.50	−146.07	−149.74	−153.50	−157.35	−53.17	
Electricity	17.00				−1.83	−37.54	−57.73	−59.18	−60.66	−62.19	−63.75	−65.35	−66.99	−68.67	−70.39	−23.79	

(continued)

Year		0	1	2	3	4	5	6	7	8	9	10	11	12	13	14	15
Processing $M @ ($/t ore)	43.00				−4.63	−94.96	−146.02	−149.69	−153.45	−157.30	−161.24	−165.29	−169.44	−173.69	−178.05	−60.17	
Service and Administration $M @ ($/t ore)	3.50				−0.38	−7.73	−11.89	−12.18	−12.49	−12.80	−13.12	−13.45	−13.79	−14.14	−14.49	−4.90	
Final Rehabilitation $M	30.00																−43.51
Annual operating expenses $M					**−25.37**	**−310.29**	**−471.46**	**−483.30**	**−495.43**	**−507.86**	**−520.61**	**−533.68**	**−547.07**	**−560.80**	**−574.88**	**−203.76**	**−43.51**
TOTAL ANNUAL OPERATING COST $M					**−26.98**	**−343.49**	**−522.64**	**−535.89**	**−549.47**	**−563.40**	**−577.69**	**−592.33**	**−607.35**	**−622.75**	**−638.54**	**−225.32**	**−43.51**
TOTAL ANNUAL OPERATING COST $/t					−269.84	−171.75	−174.21	−178.63	−183.16	−187.80	−192.56	−197.44	−202.45	−207.58	−212.85	−227.83	
Depreciation and write−offs $M			−12.49	−38.42	−52.02	−46.18	−41.59	−37.98	−35.18	−33.01	−31.35	−30.11	−29.20	−28.56	−19.65	−19.40	−77.59
OPERATING INCOME BEFORE INTEREST AND TAXES $M		0.00	−12.49	−38.42	−55.92	84.63	166.90	177.47	187.47	197.05	206.37	215.51	224.57	233.62	251.22	63.36	−35.75
OPERATING INCOME BEFORE TAXES $M		0.00	−12.49	−38.42	−55.92	84.63	166.90	177.47	187.47	197.05	206.37	215.51	224.57	233.62	251.22	63.36	−35.75
Less losses carried forward		0.00	−12.49	−50.91	−106.84	−22.21	0.00	0.00	0.00	0.00	0.00	0.00	0.00	0.00	0.00	0.00	−35.75
TAXABLE INCOME $M		0.00	0.00	0.00	0.00	0.00	144.69	177.47	187.47	197.05	206.37	215.51	224.57	233.62	251.22	63.36	0.00
Less Income Tax $M @ a rate of	30%	0.00	0.00	0.00	0.00	0.00	−43.41	−53.24	−56.24	−59.12	−61.91	−64.65	−67.37	−70.09	−75.36	−19.01	0.00
NET AFTER−TAX PROFIT $M		0.00	0.00	0.00	0.00	0.00	101.28	124.23	131.23	137.94	144.46	150.86	157.20	163.54	175.85	44.36	0.00
(NPI/FC) Less Dividends to Government as % of NA−PT @	**15.0%**	0.00	0.00	0.00	0.00	0.00	−15.19	−18.63	−19.68	−20.69	−21.67	−22.63	−23.58	−24.53	−26.38	−6.65	0.00
NET AFTER−TAX PROFIT AFTER GOVERNMENT DIVIDEND $M		0.00	0.00	0.00	0.00	0.00	86.09	105.60	111.54	117.25	122.79	128.23	133.62	139.01	149.47	37.70	0.00

Add back																	
Losses incurred		0.00	−12.49	−38.42	−55.92	0.00	0.00	0.00	0.00	0.00	0.00	0.00	0.00	0.00	0.00	0.00	−35.75
Cash effect of losses carried forward recouped			0.00	0.00	0.00	84.63	22.21	0.00	0.00	0.00	0.00	0.00	0.00	0.00	0.00	0.00	0.00
Depreciation			12.49	38.42	52.02	46.18	41.59	37.98	35.18	33.01	31.35	30.11	29.20	28.56	19.65	19.40	77.59
Initial working capital $(Year 0)M to be disbursed in Year 2	15.00			−15.76													20.95
CAPEX (calculated in separate worksheet), Total $(Year 0)M	325.00	0.00	−83.29	−256.14	0.00	0.00	0.00	0.00	0.00	0.00	0.00	0.00	0.00	0.00	0.00	0.00	0.00
Sustaining capital as % of Total CAPEX	4%	0.00	0.00	0.00	−14.00	−14.36	−14.72	−15.08	−15.46	−15.85	−16.25	−16.66	−17.08	−17.50	−17.94	−18.39	0.00
NET A−T CASH FLOW before dividend to Government as % of CF		0.00	−83.29	−271.90	−17.91	116.45	135.17	128.49	131.26	134.40	137.89	141.68	145.75	150.06	151.18	38.70	62.79
Government's receipts for royalty and NPI		0.00	0.00	0.00	1.62	33.20	66.37	71.23	73.73	76.23	78.75	81.28	83.86	86.48	90.04	28.22	0.00

Capital Expenditure and Depreciation Profile

CAPITAL COSTS															
Total Capital in $(Year 0) M	325														
Percentage spent in year 1	25%														
Percentage spent in year 2	75%														
Percentage of various capital categories Year 1 and 2, all Year 6 capital pooled															
Immediately expensed	15%														
Normal depreciable assets	25%														
Pooled project assets	60%														
Weighted average useful life of normal depreciable assets (y)	10.00														
Capital allowance for pooled assets (%)	200%														
Sustaining capital as % of Total Capital	4%														
YEAR	**0**	**1**	**2**	**3**	**4**	**5**	**6**	**7**	**8**	**9**	**10**	**11**	**12**	**13**	**14**
Year 0 (real) dollar CAPEX estimates $M															
Immediately expensible items		12.2	36.6												
Normal depreciable assets		20.3	60.9												
Pooled project assets		48.8	146.3												
Sustaining Capital				13.0	13.0	13.0	13.0	13.0	13.0	13.0	13.0	13.0	13.0	13.0	13.0
ANNUAL REAL DOLLAR CAPITAL EXPENDITURE		81.3	243.8	13.0	13.0	13.0	13.0	13.0	13.0	13.0	13.0	13.0	13.0	13.0	13.0
Inflator	1.0000	1.0200	1.0404	1.0612	1.0824	1.1041	1.1262	1.1487	1.1717	1.1951	1.2190	1.2434	1.2682	1.2936	1.3195
Real cost escalator	1.0000	1.0050	1.0100	1.0151	1.0202	1.0253	1.0304	1.0355	1.0407	1.0459	1.0511	1.0564	1.0617	1.0670	1.0723
Nominal dollar CAPEX estimates $M															
Immediately expensible items	0	12.5	38.4	–	–	–	–	–	–	–	–	–	–	–	
Normal depreciable assets	0	20.8	64.0	–	–	–	–	–	–	–	–	–	–	–	
Pooled project assets	0	50.0	153.7	–	–	–	–	–	–	–	–	–	–	–	
Sustaining Capital	0	–	–	14.0	14.4	14.7	15.1	15.5	15.9	16.2	16.7	17.1	17.5	17.9	18.4

ANNUAL NOMINAL DOLLAR CAPITAL EXPENDITURE	0	83.3	256.1	14.0	14.4	14.7	15.1	15.5	15.9	16.2	16.7	17.1	17.5	17.9	18.4
ANNUAL WRITE-OFFS AND DEPRECIATION $M															
Immediately expensible items		12.5	38.4												
Normal depreciable assets															
Average effective life (years)	10.00			8.5	8.5	8.5	8.5	8.5	8.5	8.5	8.5	8.5	8.5		
Pooled project assets															
Average effective life (years)	10.00														
Premium	200%														
Opening written down value		50.0	203.7	203.7	174.1	150.8	132.4	118.0	106.8	98.1	91.5	86.5	82.9	80.3	78.6
Sustaining capital		0.0	0.0	14.0	14.4	14.7	15.1	15.5	15.9	16.2	16.7	17.1	17.5	17.9	18.4
Pooled Project Asset Value		50.0	203.7	217.7	188.5	165.5	147.5	133.5	122.6	114.3	108.1	103.6	100.4	98.2	97.0
Depreciable Base		99.9	407.3	435.3	377.0	331.0	295.0	266.9	245.2	228.7	216.3	207.2	200.7	196.5	194.0
Depreciation pooled project assets				43.5	37.7	33.1	29.5	26.7	24.5	22.9	21.6	20.7	20.1	19.6	19.4
TOTALWRITE-OFFS AND DEPRECIATION		12.5	38.4	52.0	46.2	41.6	38.0	35.2	33.0	31.4	30.1	29.2	28.6	19.6	19.4

Cut-off Grade Calculations at Varying Royalty Rates

Royalty rate	Tonnes	Cut-off	LoM	Royalty Pd $'m	Sterilised $'m	Lost tax $'m	Welfare $'m	Sub-total F+G+H $'m	Lost income $'m
0.0%	**30.09**	2.99	10.00	–	–	–	–	–	–
0.5%	29.79	3.01	9.90	18.38	24.74	5.85	3.34	33.93	16.71
1.0%	29.49	3.02	9.80	36.57	49.97	11.76	6.72	68.44	33.59
1.5%	29.19	3.04	9.70	54.57	75.68	17.72	10.13	103.53	50.63
2.0%	28.89	3.05	9.60	72.36	101.89	23.74	13.57	139.20	67.84
2.5%	28.58	3.07	9.50	89.95	128.59	29.82	17.04	175.45	85.21
3.0%	28.34	3.08	9.42	107.54	146.29	34.84	19.91	201.03	99.54
3.5%	28.10	3.10	9.34	124.98	164.38	39.90	22.80	227.08	113.99
4.0%	27.85	3.11	9.26	142.28	182.88	45.00	25.72	253.60	128.58
4.5%	27.61	3.13	9.18	159.42	201.77	50.16	28.66	280.59	143.31
5.0%	27.36	3.14	9.09	176.41	221.06	55.36	31.63	308.06	158.17
5.5%	27.12	3.16	9.01	193.23	240.75	60.61	34.63	335.99	173.17
6.0%	26.87	3.17	8.93	209.89	260.84	65.90	37.66	364.40	188.30
6.5%	26.63	3.19	8.85	226.38	281.32	71.25	40.71	393.28	203.57
7.0%	26.39	3.20	8.77	242.69	302.20	76.64	43.79	422.63	218.97
7.5%	26.14	3.22	8.69	258.82	323.48	82.08	46.90	452.46	234.50
8.0%	25.90	3.23	8.61	274.77	345.16	87.56	50.03	482.76	250.17
8.5%	25.65	3.25	8.53	290.52	367.23	93.09	53.20	513.52	265.98
9.0%	25.41	3.26	8.44	306.09	389.71	98.67	56.38	544.76	281.92
9.5%	25.16	3.28	8.36	321.45	412.58	104.30	59.60	576.48	298.00
10.0%	24.92	3.29	8.28	336.62	435.84	109.97	62.84	608.66	314.21
10.5%	24.68	3.31	8.20	351.57	459.51	115.70	66.11	641.32	330.56
11.0%	24.43	3.32	8.12	366.32	483.57	121.46	69.41	674.45	347.04
11.5%	24.19	3.33	8.04	380.85	508.04	127.28	72.73	708.05	363.66
12.0%	23.94	3.35	7.96	395.16	532.89	133.14	76.08	742.12	380.41
12.5%	23.70	3.36	7.88	409.24	558.15	139.05	79.46	776.66	397.29
13.0%	23.45	3.38	7.79	423.09	583.81	145.01	82.86	811.68	414.31
13.5%	23.21	3.39	7.71	436.71	609.86	151.01	86.29	847.17	431.47
14.0%	22.97	3.41	7.63	450.09	636.31	157.07	89.75	883.13	448.76
14.5%	22.72	3.42	7.55	463.23	663.16	163.17	93.24	919.56	466.19
15.0%	22.48	3.44	7.47	476.12	690.40	169.31	96.75	956.46	483.75
15.5%	22.23	3.45	7.39	488.76	718.05	175.51	100.29	993.84	501.45
16.0%	21.99	3.47	7.31	501.15	746.09	181.75	103.86	1,031.69	519.28
16.5%	21.74	3.48	7.23	513.27	774.52	188.04	107.45	1,070.01	537.25
17.0%	21.50	3.50	7.15	525.12	803.36	194.37	111.07	1,108.80	555.35
17.5%	21.26	3.51	7.06	536.71	832.60	200.75	114.72	1,148.07	573.58
18.0%	21.01	3.53	6.98	548.02	862.23	207.18	118.39	1,187.80	591.96
18.5%	20.77	3.54	6.90	559.05	892.26	213.66	122.09	1,228.01	610.46
19.0%	20.52	3.56	6.82	569.80	922.68	220.19	125.82	1,268.69	629.10
19.5%	20.28	3.57	6.74	580.26	953.51	226.76	129.58	1,309.84	647.88
20.0%	20.03	3.59	6.66	590.43	984.73	233.38	133.36	1,351.47	666.79

Appendix E: OECD Guidelines' Five Transfer Price Estimation Methods (Reproduced from Guj et al. 2017)

The OECD *Transfer Pricing Guidelines for Multinational Enterprises and Tax Administrations, 2012* present five methods which are used to apply the arm's length principle including**:**

- **Three** traditional methods **i.e.:**

  The **Comparable uncontrolled price (CUP)** method directly compares the price charged for property or services transferred in a controlled transaction to that charged in a comparable uncontrolled transaction under comparable circumstances where available.

  Adjustments may be needed to achieve comparability and five comparability factors[1] are identified in this regard. The reliability of the CUP method is affected by the degree of accuracy of the adjustments made.

  The **Resale price method (RSP)** derives the price at which a product has been purchased from an associated enterprise by netting from the price at which it has been subsequently resold to an independent buyer an appropriate gross margin. This margin represents the amount which the reseller would seek to cover its selling and other operating expenses and, in light of the functions performed (taking into account assets used and risk assumed), make an appropriate profit. This method is most useful where applied to marketing operations.

  The **Cost plus (CP)** method applies an appropriate mark-up to the costs incurred by the supplier of property or services to an associated purchaser in a controlled transaction. The mark up represents an appropriate profit in light of the functions performed and the market conditions. This method is most useful where semi-finished goods are transferred or services provided between associated parties, under joint facility or long-term supply arrangements.
- and a further two **transactional profit methods** i.e.:

  The **Transactional net margin method (TNMM)** examines the net profit that a taxpayer realises from a controlled transaction or aggregate transactions relative to an appropriate base (e.g. costs, sales, assets). It operates in a manner similar to the cost plus and resale price methods but requires the selection of the most appropriate net profit indicator and its weighting. TNMM is suitable to cases where one of the parties makes all/most the unique contributions involved in the controlled transaction.

---

[1]Main comparability factors:

I. Characteristic of the property or services transferred
II. Functions performed by the parties (taking into account assets used and risks assumed)
III. Contractual terms
IV. Economic circumstances of the parties
V. Business strategies pursued by the parties.

The **Transactional profit split method (PSM)** determines the division of the profit from a controlled transaction between associated enterprises on the basis of what independent enterprises would have agreed at arm's length and expected to realise from engaging in the same transaction. This method can offer a solution for highly integrated operations for which a one-sided method would not be appropriate and also be most appropriate where both parties make unique and valuable contributions to a transaction.

In practice, the level of information available is often inadequate to conduct a satisfactory comparability analysis and some flexibility and the exercise of good judgment become necessary in reaching a reliable arm's length estimate.

The OECD[2] has also considered, at the request of some developing nations, the use of the so-called **'sixth method'** which makes mandatory use of publicly quoted prices for commodities on their shipment date to a related party. This method, initially employed in Argentina, has now been adopted by most South American countries and by India[3].

[2]OECD (2014), Transfer pricing comparability data and developing countries.

[3]PriceWaterhouseCoopers, PKN Alert, January 29, 2013.

Appendix F: BEPS Action Plan

Action 1: Addressing the Tax Challenges of the **Digital Economy**
Action 2: Neutralising the Effects of **Hybrid Mismatch Arrangements**
Action 3: Designing Effective **Controlled Foreign Company Rules**
Action 4: Limiting Base Erosion Involving **Interest Deductions** and Other Financial Payments
Action 5: Countering **Harmful Tax Practices** More Effectively, Taking into Account Transparency and Substance
Action 6: Preventing the Granting of **Treaty Benefits** in Inappropriate Circumstances
Action 7: Preventing the Artificial Avoidance of **Permanent Establishment** Status
Actions 8–10: Aligning **Transfer Pricing** Outcomes with Value Creation
Action 11: Measuring and Monitoring BEPS
Action 12: Mandatory **Disclosure Rules**
Action 13: Guidance on **Transfer Pricing Documentation** and Country-by-Country Reporting
Action 14: Making **Dispute Resolution** Mechanisms More Effective
Action 15: Developing a **Multilateral Instrument** to Modify Bilateral Tax Treaties

Glossary

Advance Pricing Agreements Is an arrangement between a taxpayer and the taxation authority whereby the transfer prices to be applied to future intra-group transactions between the taxpayer and related parties are determined in advance and remain valid over a number of years after which they are subject to review.

Allocative Efficiency Is achieved when resources are being re-allocated to their most productive use to generate the ever changing mix of goods and services that society requires.

Arms-length principle Requires that the transfer price adopted in transactions between related parties should be the same that would have been negotiated for the same goods and services by independent parties without compulsion in a contestable market.

BEPS Action Plan Was developed by the OECD and supported by all its members and the majority of other jurisdictions to prevent erosion of the tax base of and shifting of profits from developing countries. The plan which includes 15 specific actions was finalised in 2015 and is currently being progressively implemented through appropriate amendments of the domestic tax laws of various jurisdictions.

Brownfield Refers to an exploration terrane in which some mineral deposits have already been discovered and developed. It generally features reasonably high probability of future discoveries albeit of more moderate size than those initially expected in greenfield exploration.

Build, own, operate and transfer (BOOT) An arrangement sometimes used to fund public infrastructure based on a government-private enterprise partnership whereby the latter funds, builds, owns and operates the facilities and sometimes transfer back to government their ownership.

Capital allowance Provides a tax-paying entity with rules governing how much expended capital can be offset against taxable income over a specific period. An allowance in excess of 100% may be used by government as an incentive for industry to invest in new assets.

E. Lilford, P. Guj, *Mining Taxation*, Modern Approaches in Solid Earth Sciences 18,
https://doi.org/10.1007/978-3-030-49821-4

Capital gain tax Is the tax levied on the difference between the price realised in the sale of a capital asset and its written down value in the book of the seller, assuming that the item is sold at a profit. Otherwise a capital loss would be incurred which, in most jurisdictions may only be offset for tax purposes against future capital gains rather than operating taxable income.

Comparables Is a term denoting information about the price realised for various goods and services in transactions between unrelated parties in contestable markets which is commonly used in comparison with transfer prices used in transactions between related parties to ascertain whether they were at harm-length.

Comparative advantage Is achieved if a country specializes in producing and exporting primarily those goods and services which it can produce more efficiently and relies on importing goods and services which it cannot produce competitively.

Contract of work Is a legally binding agreement entered into between two parties wherein one party is being contracted to provide specific services or products at a specified cost for deliverables agreed to with the counter-signing party. In some jurisdictions, typically Indonesia, it often takes the form of a standard mining contract, entered between government and industry, specifying all the conditions, including those pertaining to mineral royalties and taxation, under which a mining project can be developed and operated.

Cost-benefit analysis A form of economic project evaluation which includes both monetary and non-monetary benefits and costs the value of which is estimated using a methodology called 'shadow pricing' and discounted at the 'social' rate of discount.

Country-by-Country (CbC) Reporting Is an element of the mandatory **disclosure rules** under BEPS Actions 12-13 that require stringent transfer pricing documentation and reporting on a project-by-project, government-by-government and country-by-country basis.

Cut-off grade Is the amount of concentration of a mineral or metal in a deposit, typically expressed as either a percentage or, in the case of precious metals as grams per tonne, such that if exploitation of that deposit occurred and the operations achieved a grade equivalent to the cut-off grade, the project would break even (zero profit or loss). The cut-off grade which must be calculated with consideration of operating costs, capital costs, commodity prices and exchange rate forecasts, determines the tonnage of mineable reserves and as a consequence the life of a mine.

Depletion allowance Is a fiscal incentive to invest in mining created by the deductibility of the value of exploited reserves for the purpose of determining the taxable income of a mining operation. It is justified on the basis that the orebody reserves are non-renewable.

Diluted mining reserves Represent the amount of ore that is economically extractable at a pre-determined mining rate, cost and under specified market condition. During ore extraction some waste or low-grade material will inevitably be mixed up with the undiluted ore reserves diluting them. Dilution has a negative impact on the performance of a project as it increases the tonnages to be transported and reduces their average grade and therefore value.

Diminishing value or declining balance depreciation Is an accelerated depreciation method that generates larger depreciation amounts over the earlier years of an operation, and smaller depreciation amounts over the later years of that operation. Declining balance depreciation is calculated as the product of the depreciable (current book) value of an asset and the depreciation rate. This method accommodates rapidly devaluing assets or assets that become obsolete over a short period of time. Alternatively it may be adopted by government as a fiscal incentive and to lessen the taxation burden in the early more vulnerable years of a mining operation.

Distributional Efficiency Is achieved when the composition of the output of an economy is such that consumers would not wish, given their disposable income and market prices, to spend it in any different way.

Dore' Is an amalgam of gold and various amounts of silver. Dore' will need to be further refined to extract the gold and the silver separately into high-purity ingots.

Double taxation agreement Is an agreement between two countries designed to avoid the same profit being taxed twice in each country. The country to which profit after tax are repatriated generally will only impose a tax on the difference between the taxable income as assessed by them and that assessed in the country where the profit was first generated. Double taxation agreement typically also moderate the reciprocal rates of withholding taxes and may include arrangements for the exchange of information needed in audits. Multilateral tax treaties may also be negotiated along the same lines as those applying to bilateral agreements among more than two countries.

Dutch disease Is an economic term that explains how when a specific sector in a country, such as natural resources, booms, that is to say performs better than most other sectors, it may attract a significant share of employment into that sector paying higher wages and salaries to the detriment to the other lagging sectors in the same country, which as a consequence would experience an under-supply of labour.

Economic efficiency Is a state of the economy in which every resource is optimally allocated to serve each individual or entity in the best way while minimizing waste and inefficiency.

Economic externalities Are costs or benefits, often non-monetary, incurred or received by a third party who has no control over their creation. An **externality** can be either positive or negative and can be generated by either the production or consumption of a good or service.

Economic rent based royalty A system where the royalty payment is a percentage of the economic rent generated by a project computed on a cash basis after allowing a measure of 'normal' profit.

Economic rent Is the cash profit generated at a project level after deducting from revenue all capital and operating costs of production including a measure of 'normal' profit that is to say a return on investment adequate to compensate investors for bearing the risk of the project.

Employment multiplier Is a measure of the indirect employment impact that a particular industry will have throughout the economy of the region/country in which it operates. It is generally presented in terms of the number of indirect jobs that may be created in an area for each direct job created in a specific industry.

Engineering useful asset life Is the expected and hence planned for duration that a specific asset or piece of plant or equipment is likely to be able to be gainfully utilised in an operation before having to be replaced because of likely increased mechanical inefficiency and reduced reliability. It provides a guide as to when it would be more advantageous to replace an existing piece of equipment with a new one rather than continuing to maintain it. The engineering useful asset life is different from the fiscal useful asset life and related depreciation rates which are provided by the tax authority to establish the related depreciation deductions used in determining the taxable income of a project.

Extractives Dependence Index As developed by the UNDP includes three basic indicators: (1) share of total export earnings from extractive activities; (2) revenue from extractive activities as a share of total fiscal revenue; and (3) value added by extractive activities as a percentage of the total.

Fiscal incentive A tax policy designed to engender a desired taxpayer's behaviour generally by reducing the impact of taxes on him/her.

Fiscal policy Focuses on taxation levels and government spending.

Fiscal useful asset life Is the set by the tax authority for the purpose of calculating the amount of depreciation to be deducted in calculating the taxable income of a project. Assets are generally predetermined to have fiscal useful lives shorter than their mechanical or engineering ones and, as a consequence, they may continue to be utilised in practice in operations beyond the time when they have been fully depreciated for tax purposes.

Flow-through share Is a Canadian taxation incentive to attract investment funding into the mineral exploration sector. Investors in an exploration company that is predominantly exploring within Canada are allowed to offset the expenditure incurred in exploration by the company over each period against their personal taxable income, thus deriving an immediate tax benefit from their investment.

Free carried interest An equity interest at the local holding-company, project or asset level.

Fringe benefit tax A tax imposed on benefits bestowed by a company on its employees other than by means of conventional wage or salary remuneration. It may include the value of subsidised accommodation or meals in a mining town as well as other benefits.

Fuel excise Is a duty (tax) payable to the government on the purchase (import) of fuels. The excise is charged at a predetermined rate which is either a percentage of the price of the fuel or a set amount per litre.

Grade-tonnage trade-off curve Is a graph showing what the amount of ore exists above any predetermined cut-off grade. This knowledge allows the mine planners to determine an optimal trade-off between grade and tonnage of ore that could commercially be mined by various methods to achieve specific economic outcomes.

Grand parenting Is a colloquialism to denote allowing some interested parties to be quarantined from policy changes which would otherwise have deprived them of some existing privileges or asset values.

Greenfield A potential exploration property or area on which no exploration or extrapolation has taken place yet. The area is likely to contain mineralisation but no work has been completed as yet for this to be confirmed.

High-grading Where the operator or owner of the project elects to specifically mine the higher grade ore at the expense of blending with, or mining of, lower grade material. The higher grade ore is preferentially depleted before any lower grade ore is mined.

Horizontal fiscal equalisation policy Is designed to ensure that the standards of living in the various states/provinces of a federation are maintained to a comparable level by redirecting revenues raised in, for instance mineral-rich states, to less endowed states. In Australia this is achieved by a redistribution of GST revenues in a process conducted every 4 years by the Grant Commission.

Information Exchange Agreement Are a common component of bilateral and multilateral double taxation agreements to ensure that information necessary in the conduct of audits of cross-border transactions can be easily and timely obtained.

IONEX Index Is a benchmark assessment provided by Platts of the daily spot price of physical **iron ore** based on a standard specification of **iron ore** fines with 62% **iron,** 2.25% alumina, 4% silica and 0.09% phosphorus, among other gangue elements. Indices for other iron grades (e.g. 58%) are also published. This type of index is commonly used to revise, frequently on a quarterly basis average, the price applicable in long-term offtake agreements.

Licence to operate In the context of an actual permit, this is a government awarded license to physically extract and treat or sell ore. A social license to operate differs in that no actual permit exists, but is rather a mutual agreement for mining activities to take place with the support of the local or regional community or other affected parties.

Life of mine Is the calculated duration of mining operating activities likely to occur before the full depletion of the economic part of the finite ore body is realised.

Limitation of benefits test Determines whether a taxable entity has an adequately substantial presence in the territory of residence to justify local taxation of profits which have been declared in that jurisdiction.

Mineral economy That part of a wider economy that is directly related to and impacted by the minerals sector alone. If the output of the minerals sector represents a large proportion of the GDP of a state/nation, typically more than 8%, the latter is generally classed as a mineral economy. The public funds and currency of a mineral economy tend to fluctuate as a reflection of the volatility of commodity markets.

Mineral exploration incentive scheme Is generally a financial incentive mechanism put in place that encourages mineral exploration activities to occur. The flow-through share mechanism is an example of a mineral exploration incentive, but other examples may include non-tax benefits as the incentive.

Misinvoicing Is the deliberate and fraudulent practice of avoiding taxes in the country hosting mining operations by shifting profits to a related entity resident in a low tax jurisdiction by understating the tonnages and above all the grades of mineral exports sold to it.

Monetary policy Focuses on the central bank functions of setting the money supply and interest rates.

Multilateral instrument Is an arrangement under BEPS Action 15 allowing governments to modify existing bilateral tax treaties in a synchronised and efficient manner to quickly implement treaty-related BEPS recommendations without the need to expend resources renegotiating each treaty bilaterally. To date over 100 jurisdictions have agreed to participate in this initiative.

Net smelter return Is the revenue received by the project owner (being a percentage of the income from sales or a royalty rate) for the sales of the mined product, less any costs and less any on-mine and off-mine transportation and beneficiating costs.

Net smelter value Is the quantification of income (revenue) received, being a reflection or determinant of value, by a project owner after the sales of the product less transport and beneficiating costs incurred and less the smelter costs.

Non-participative (non-contributory) interest See free carried interest.

Normal profit A return on investment adequate to attract and retain investment funds in a project with cognisance of its risk.

Off-take agreement Is the agreement entered into by the owner of a mineral project and either the end-user or an intermediate user of the produced material for further refinement, wherein the terms measured by quantity, grade (purity) and other factors are agreed between the two parties, and the purchaser guarantees to acquire all or an agreed amount of that product at an agreed price or at the spot price.

Permanent Establishment Is the place where one or more expatriates carry out business in a country often intermittently but generally over relatively long aggregated periods of time.

Principal purpose test Is carried out to prove that the motivation for establishing a business in the jurisdiction was not primarily to derive tax benefits from the presence of a double taxation agreement.

Production sharing contract Is similar to a royalty other than it is the value of a specified quantity of production that is provided to the holder of the contract. In rare circumstances, a production sharing contract may result in the physical sharing of production and not an actual monetary payment.

Productive Efficiency Is achieved when the output of the economy is being produced at the lowest unit cost.

Profit-based royalty A system where the royalty payment represents a percentage of the profit generated by the project. The profit base is generally calculated according to specific provisions of the relevant royalty regulations which may be at variance with those of the CIT legislation.

Progressive ad valorem royalty Is a system where the royalty rate increases in pre-determined steps as the price of the mineral commodity increases.

Resource rent tax See Economic rent based royalty.

Ring fencing Is a tax practice that requires that individual mining projects be held by a corporate taxable entity specifically registered for this purpose in the country. A foreign resident MNE can be the whole or controlling shareholder of such a company.

Secondary Tax on Companies This additional tax, which has been replaced by a dividends tax in many of the jurisdictions in which it has been imposed, taxes the after tax profits of a company when that company looks to make a distribution to shareholders.

Service centres or hubs Are set up by MNEs to provide goods and services to their regional or global operations. The location of a hub can be selected to achieve logistical or operational synergies, but more often to minimise tax at the consolidated MNE level.

Special (stability) agreements Is an agreement between an operator and a government whereby the government guarantees fiscal stability over a finite period, generally over the life of the operation. The fiscal stability will typically cover taxes, royalties, duties and other policies that may be subject to change, especially as a consequence of a change in government parties.

Special purpose vehicle/entity Is a company's wholly-owned subsidiary specifically created to isolate financial risk for activities that may attract greater risk. It is also a separate, ring-fenced tax entity. The vehicle or entity may house certain assets or operations for structuring benefits including securing equity and debt, and for tax efficiency purposes.

Specific or unit-based royalty A system where a specified amount is levied on each cubic metre or, more frequently tonne of ore mined or sold.

Sterilization of resources If, due to political or economic impositions, the cut-off grade (see above) is forced to increase so that the operation may still break even, additional unmined mineral will result as the operation targets the higher grades. This additional material that was previously economic but is rendered uneconomic due to the policy changes will be permanently unprofitable and hence sterilized.

Strait line depreciation Is a depreciation method whereby the asset value decreases at a constant rate over time. The asset may be depreciated fully over its useful life or a residual value estimated and deducted from the depreciable asset value.

Tax holiday Is a tax incentive in the form of a period, frequently up to the first 5 years of production, during which a mining operation will pay no corporate income tax and in some cases no mineral royalties.

Tax rate arbitrage Is the MNEs' strategy of reducing their tax liability at the consolidated level by shifting profits to low or no-tax jurisdictions.

Taxing point Is the point along the mining cycle value chain at which the royalty value base is set. This is typically at the smelter, fob the port of export or at the mine gate.

Thin capitalisation rule Is a tax rule that limits the ratio of debt to equity that can be used in the funding structure of a mining company, typically between 2:1 and 3:1.

Transfer price Is the price adopted in controlled transactions between related parties which are part of the same MNE group.

Treaty shopping A colloquial term to denote the MNEs' strategy of determining their global structures to minimise their tax at the consolidated level by registering some of their service providing hubs in low or no tax jurisdictions which have bilateral or multilateral tax treaties with the highly taxing country hosting the tax mining operations, which have the effect of reducing or eliminating withholding tax payments.

Value added or goods and services tax Is an indirect tax applied as a percentage, typically 5–20%, to the value of the goods and services sold to a third party where value has been added to them.

Value-based or *ad valorem* royalty A system where the royalty payment is derived by applying a percentage rate to the realised value of sales of mineral products.

Withholding tax Is deducted as a percentage of payments remitted abroad for dividends, interest and for the services of expatriate specialists.

The manufacturer's authorised representative in the EU is Springer Nature Customer Service Centre GmbH, Europaplatz 3, 69115 Heidelberg, Germany. If you have any concerns regarding our products, please contact ProductSafety@springernature.com

Printed and bound by CPI Group (UK) Ltd, Croydon, CR0 4YY
30/09/2026
02225505-0008